The Hitchhiker's Guide to Calculus

A Calculus Course Companion

Other books by Michael Spivak

Calculus, 3rd ed.
  Publish or Perish, Inc.

Calculus on Manifolds
  Benjamin/Cummings

A Comprehensive Introduction to Differential Geometry (5 volumes)
  Publish or Perish, Inc.

The Joy of T$_E$X
  American Mathematical Society

The Hitchhiker's Guide to Calculus

A Calculus Course Companion

Michael Spivak

Published and Distributed by
The Mathematical Association of America

This edition published by
The Mathematical Association of America (Incorporated)
Washington, DC

ISBN: 0-88385-812-6
Library of Congress Catalog Card Number: 2003113206

Printed in the United States of America

Current printing (last digit):
10 9 8 7 6 5 4 3 2 1

The Hitchhiker's Guide to Calculus

1 *Where are we going? An opportunity for the reader and the author to get acquainted.*

Y ou never know when a little esoteric knowledge will come in handy. The summer that I hitchhiked through Europe finally found me in Italy, where I took a bus to the Adriatic coast to bake on one of the tiny beaches nestled in the cliffs, before taking a train for a plane home from Rome. The train ride was long, delayed, and hot, and as we neared the station I felt a sharp pain in my left side. Recalling enough elementary biology to realize it couldn't be my appendix, I figured that dehydration had probably just produced a cramp.

Still, ... , the pain didn't abate, and it was going to be a long flight home, so I dragged myself and my backpack to the nearest hospital. I spoke not a word of Italian, and the doctors spoke scarcely any more English, but I knew that such difficulties weren't going to deter them from trying to be helpful. The doctor kept groping for a word that I would comprehend, until he tentatively offered *calculi*. Ah! ... I understood—perhaps I had kidney stones.

Well, it eventually turned out that I didn't have kidney stones, or anything else interesting, for that matter—it was probably just a cramp all along. But at the time it was certainly comforting to have some sort of diagnosis. Fortunately, I knew that a *calculus* was a stone, or pebble, because I knew how the branch of mathematics called "Calculus" got its name.

Little stones were once used for calculations like addition and multiplication, so any new method of calculating came to be called a "calculus." For example, when methods for calculating probabilities were first introduced, they were called the Probability Calculus. Nowadays, we would just call it probability theory, and the word "calculus" has similarly disappeared from most other such titles. But one new method of calculation, The Differential and Integral Calculus, was so important that it soon became known as *The* Calculus, and eventually simply Calculus.

The point of this little story is not to supply one more bit of esoteric knowledge (or perhaps aid with a hospital diagnosis in a foreign land), but to emphasize a point about Calculus, that mysterious, and perhaps forbidding sounding, new branch of mathematics.

Every new method of calculation involves some new ideas. To learn the rules for algebraic calculations you first have to absorb the idea of using letters for unknown quantities; for trigonometry you have to get used to the idea of using similar triangles to attach numbers to angles. Similarly, Calculus requires understanding a whole new set of ideas, ones which are very interesting and quite beautiful, but admittedly also a bit hard to grasp. Yet, with all the new ideas that it entails, Calculus *is* a method of calculation, so in your Calculus course you are going to be doing calculations, reams of calculations, oodles of calculations, a seeming endless number of calculations!

There's really no way to avoid this. The only way to understand the new ideas is to calculate with them until the calculations can be done easily and almost without thought. The trouble is, with all the time spent on these calculations, it becomes easy to lose sight of the basic ideas behind them. After a while one can't see the forest for the trees, or—perhaps it would be more appropriate to say—the beach for the pebbles.

So the aim of this guide is to provide a carefree jaunt that explores the new ideas of Calculus. Occasionally we might actually do a calculation or two, just to make sure that we really understand what we have seen, but our emphasis will be on the highlights—there will be time enough in your courses for the dirty work. We're going to be traveling light, and visiting all the main attractions, without getting bogged down by the daily cares that bedevil travellers encumbered with excess baggage.

2

We pack the essentials,
before embarking on our journey.

T here are all sorts of mathematical prerequisites for calculus—at some time or other, most of the techniques of algebra, trigonometry, and analytic geometry come into play. But the main thing we will need on our journey are a few of the simplest ideas from analytic geometry.

In analytic geometry we introduce a **coordinate system**, by drawing two perpendicular lines, the two **axes**, along which we measure distances; conventionally, positive numbers represent distances to the right on the horizontal axis, and distances upwards on the vertical axis.

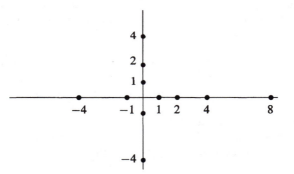

Then we associate a *pair* of numbers (a, b) to each point in the plane according to the scheme below, with a indicating horizontal distance and b indicating vertical distance. The point corresponding to $(0, 0)$ is the **origin** O, the intersection of our two axes.

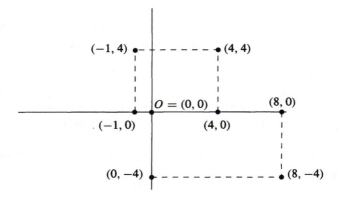

Once we have this method of associating points to pairs of numbers, and *vice versa*, we can then look at *curves* that correspond to *equations*. For example, suppose we take about the simplest curve, a straight line *L*

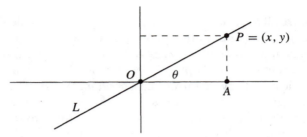

through the origin. If $P = (x, y)$ is any point on this line, then the line segment OA has length x, while the segment AP has length y. Consequently, the ratio

$$\frac{y}{x} = \frac{AP}{OA}$$

(here AP stands for the length of the segment from A to P, and similarly for OA)

is always the same number m, no matter which point P we choose: this ratio is just the tangent of the angle θ that our straight line makes with the horizontal axis; we also call it the **slope** of the line L.

Thus, any point (x, y) on L satisfies

$$\frac{y}{x} = m,$$

or the equivalent equation

(1) $$y = mx$$

(which has the advantage that it makes sense, and is true, even for $x = 0$). We express this by saying that (1) is the **equation of the straight line** L.

Because we tend to use (x, y) as the name of an arbitrary point (and therefore look for equations between x and y), the number x usually represents a distance along the horizontal axis, while the number y usually represents a distance along the vertical axis. Consequently, the horizontal axis is often called the **x-axis**, while the vertical axis is often called the **y-axis**.

Notice that one case of (1) is special: if $m = 0$, then we simply have the "equation"

$$y = 0,$$

which doesn't contain x at all. But it is still true that this "equation" represents a straight line—it is simply the equation for the horizontal axis, since the point (x, y) satisfies $y = 0$ precisely when (x, y) lies on the horizontal axis; in other words, on the x-axis, the value of x is arbitrary, while y must be 0.

How about a straight line L that doesn't go through the origin? One simple way to find the equation of L is by drawing the line L' that is parallel to L but which does pass through the origin.

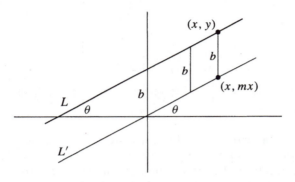

Then each point of L lies directly over a point of L', and the distance between these two points is always the same number b. Consequently, corresponding to each point (x, mx) of L', we have a point (x, y) on L with

(2) $$y = mx + b.$$

Thus (2) is the equation of our straight line L. In this equation, m is again called the **slope** of L (it is still the tangent of the angle θ that L makes with the horizontal axis). The number b is called the **y-intercept**, since it is the distance at which the line L intercepts the y-axis, and equation (2) is called the **slope-intercept** form of an equation for a straight line.

Once again, the case $m = 0$ is special: we obtain the "equation"

$$y = b,$$

which simply represents a line parallel to the x-axis.

It looks as if (2) is the most general equation we need to describe a straight line, but we've actually missed a whole batch, namely the straight lines that are vertical. However, it's easy to see that they simply have "equations" of the form

$$x = a,$$

the "equation" of a vertical straight line intersecting the x-axis at $(a, 0)$.

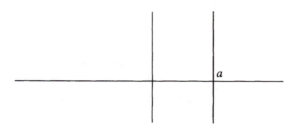

A more telling criticism of the slope-intercept equation is that it is of use only when we happen to be describing a line in terms of its slope and y-intercept, which might not be the data of interest to us. For example, the first question that might come to mind is how to find the equation of a straight line between two given points P and Q, where we are given neither the slope nor the y-intercept.

Instead of answering this question directly, we'll first consider yet another question, which will turn out to be quite useful later on: How do we find the equation of a straight line which passes through a given point $P = (x_0, y_0)$, and which has a given slope m?

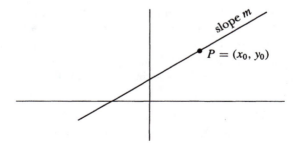

There's a perfectly straightforward, and unimaginative, way to solve this problem: Since the slope is m, we know that the straight line has an equation of the form

(a) $$y = mx + b$$

so we just have to figure out what b is. And to do this we just have to use the fact that $P = (x_0, y_0)$ lies on the line: this simply means that (x_0, y_0) satisfies (a), and thus that

$$y_0 = mx_0 + b.$$

We can simply solve this for b,

$$b = y_0 - mx_0,$$

so that (a) becomes

(b) $$y = mx + [y_0 - mx_0].$$

This is perfectly correct, but not especially appealing, nor easy to remember. More geometric reasoning easily remedies both these defects: In the figure below, the vertical segment has length $y - y_0$ and the

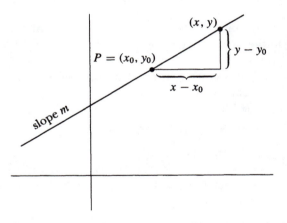

horizontal segment has length $x - x_0$. Since the line has slope m, this means that

$$m = \frac{y - y_0}{x - x_0},$$

or

(3) $$y - y_0 = m(x - x_0).$$

This equation is called the ***point-slope*** form of an equation for a straight line. It is easily seen to be equivalent to (b), but it exhibits the relationship between the given data much more clearly.

Limbering Up

Two exercises to prepare us for the rigors of the journey ahead.

1. We conveniently finessed the problem of finding the equation of the straight line through two given points $P = (x_0, y_0)$ and $Q = (x_1, y_1)$, so that it could be left as an interesting exercise.

    (a) We are looking for an equation of the form

    $$y = mx + b,$$

    where m and b are presently unknown. What we do know is that P and Q lie on this line, which means that we have the two equations

    $$y_0 = mx_0 + b$$
    $$y_1 = mx_1 + b.$$

    Solve these two equations for m and b, to obtain the desired equation. [Answers at the bottom of page 9.] (Remember that x_0, x_1, y_0 and y_1 are simply four known numbers—so this is just a simple problem involving two equations in the two unknowns m and b.)

    (b) Instead of this straightforward, unimaginative, approach, with its not very easily remembered result, we'll now try for a more elegant solution.

    Answer the following two questions:

    (i) What must the slope m be?
    (ii) Now apply the *point-slope* form, using this value of m and the fact that the line goes through P, to obtain the desired equation.

    (c) Of course we could also use the point-slope form together with the fact that the line goes through Q. Check that the equation you obtain in this way agrees with that obtained in part (b).

2. This second exercise is basically just a joke, though it's the sort of joke that only a mathematician would find amusing. Nevertheless, it has a serious point, and there's a reason why it follows the previous exercise.

    (a) You are in a room with a gas stove, one that is working, and already lit, together with a table, and a pot of water on the table. You need some boiling water. What do you do? This part isn't meant to be tricky—just give the obvious answer.

(b) Now you are in the same room, with the same gas stove, still working, and already lit, together with the same table, and the same pot of water, but the pot of water is on the floor. You need some boiling water. What do you do? This part *is* meant to be tricky—you're supposed to give an "elegant" answer rather than the obvious one.

Answers.

Problem 1(a). Subtracting the second equation from the first immediately eliminates b, giving a formula for m. Multiplying the first equation by x_1 and the second by x_0 and then subtracting, gives a formula for b.

1(b). The slope m clearly must be $y_1 - y_0/x_1 - x_0$, using the same reasoning as on page 7 [and as we computed in part (a)]. Then the point-slope form gives

$$y - y_0 = \frac{y_1 - y_0}{x_1 - x_0}(x - x_0).$$

1(c). We also get

$$y - y_1 = \frac{y_0 - y_1}{x_0 - x_1}(x - x_1),$$

which is found to be equivalent, after multiplying everything out.

Problem 2(a). Put the pot of water on the stove.

2(b). Put the pot of water *on the table*. (Because now you've reduced the problem to one that you already know how to solve, in just the same way that Exercise 1 reduced the problem of finding the equation of a line through two points to the point-slope form, which is the answer to a different problem.)

3 *We leave the straight and narrow.*
Our first encounter with something new.

In the previous chapter we worked almost exclusively with equations of the form

$$y = ax + b$$

(although we usually used m instead of a). Because the graphs of these equations are always straight lines, the expression $ax + b$ is usually called a **linear** expression.

Anything that isn't linear is called **non-linear**, but as usually happens when we try to create such dichotomies ("There are two sorts of people in this word ... "), the groups determined by this criterion are hardly comparable in size. As soon as we consider equations involving non-linear expressions we encounter an almost bewildering array of possibilities.

Let's begin with the simplest of all,

$$y = x^2.$$

If we plot lots of points (x, y) satisfying this equation, i.e., lots of points of the form (x, x^2), we get a picture like

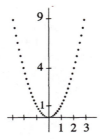

which can be filled in to give a curve called a **parabola**.

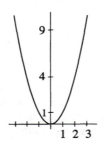

This curve is symmetric about the y-axis, since $(-x)^2 = x^2$. Moreover, it never goes below the x-axis, since we always have $x^2 \geq 0$. Similarly, the equation

$$y = -x^2$$

will describe a curve which instead never goes above the x-axis.

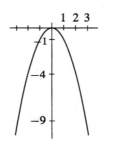

On the other hand, the equation $y = 2x^2$ describes a curve that is twice as "steep" as our first parabola, while $y = \frac{1}{2}x^2$ describes a curve that is half as steep.

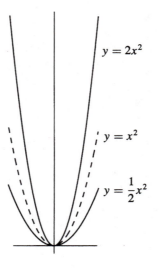

And, in general, $y = cx^2$ will describe a curve that is similar to our first parabola for $c > 0$ but similar to the second for $c < 0$.

Things look quite different for the equation $y = x^3$. This is the equation

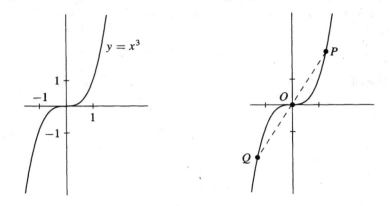

of a curve that continually increases through both negative and positive values. Corresponding to any point $P = (x, x^3)$ on this curve, the point $Q = (-x, (-x)^3)$ is also on the curve, and this point

$$Q = (-x, (-x)^3)$$
$$= (-x, -x^3),$$

is on the opposite side of the line from O to P. Thus this curve is "symmetric with respect to the origin."

This naturally leads us to consider $y = x^4$, $y = x^5$, ..., and each produces a distinctive curve. However, in general, the curves described by $y = x^n$ for even n all resemble the parabola, while the curves described by $y = x^n$ for odd n all look somewhat like the one for $n = 3$.

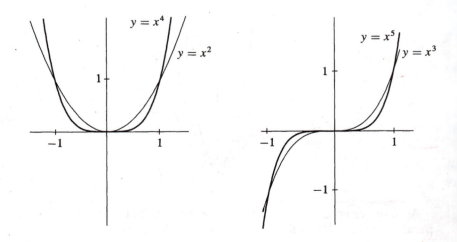

Instead of considering curves determined by equations, we often refer instead to graphs determined by functions. A ***function*** f is just a rule for assigning to any number x another number, which is denoted by $f(x)$; this number is called the "value" of f at x, and $f(x)$ is simply pronounced "f of x". By the ***graph*** of the function f we mean the collection of all points $(x, f(x))$. For example, if f is the function determined by the rule $f(x) = 3x^2 + 4$, then its graph consists of all points $(x, 3x^2 + 4)$.

We could just as well describe this as the collection of points (x, y) satisfying $y = f(x) = 3x^2 + 4$, so the graph of f is the same curve that we have previously described in terms of the equation $y = 3x^2 + 4$. Thus it might seem that speaking of "functions" rather than "equations" is just a change of vocabulary, but there are some important distinctions to be made.

First of all, a function might be a rule that cannot be given by an equation. For example, suppose that we consider the function f given by the rule

$$f(x) = \begin{cases} x & \text{if } x \geq 0 \\ -x & \text{if } x \leq 0 \end{cases}$$

with the graph shown below; notice that we always have $f(x) \geq 0$, by its definition, so that the graph never goes below the x-axis. Also the graph is symmetric with respect to the y-axis, since $f(-x) = f(x)$.

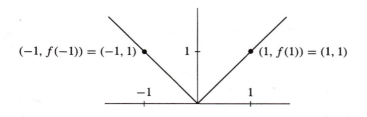

The symbol $|x|$ (the ***absolute value*** of x) is defined as

$$|x| = \begin{cases} x & \text{if } x \geq 0 \\ -x & \text{if } x \leq 0. \end{cases}$$

Of course, that means that our graph *is* the graph of an equation after all, namely the equation $y = |x|$! But obviously that's only because we decided to introduce a new symbol with which to write equations. There are many functions f whose graphs would be pretty hard to describe as the graph of an equation in any natural way. For example, the picture below shows the graph of a function f where the rule for $f(x)$ is a complicated mixture of different formulas.

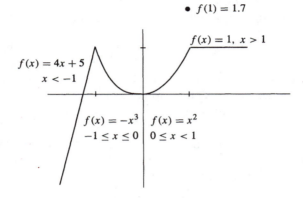

Although graphs of functions thus allow many creatures that would be hard to describe as graphs of equations, sometimes things work out the other way. For example, the graph of the equation

$$x^2 + y^2 = 1$$

is just a circle of radius 1 around the origin O, since $\sqrt{x^2 + y^2}$ is just the

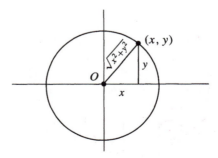

distance from $O = (0, 0)$ to (x, y). This graph *can't* be the graph of a function, because the graph of a function consists of points of the form

$(x, f(x))$, where $f(x)$ can only be one number—the graph of a function can't have two different points on the same vertical line.

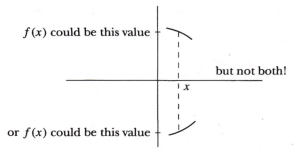

On the other hand, we can define two different functions

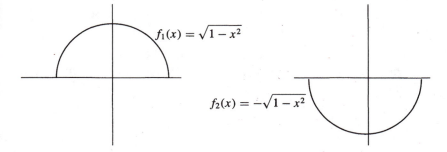

whose graphs together give a circle. In fact, we usually have to resort to tricks like this whenever we want to use calculus to investigate curves like a circle, because calculus is concerned almost entirely with analyzing functions and their graphs.

A vertical line fails to be the graph of a function in the worst possible way, but a horizontal line is simply the graph of a function defined by the

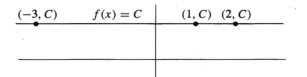

rule $f(x) = C$ for all x. Such functions are called **constant** functions.

Although not all functions can be described by equations, it's still true that a tremendous variety of functions can be described simply by giving a formula for $f(x)$, like $f(x) = x^2$ or $f(x) = x^3$. Functions are usually described briefly in just this way, instead of saying cumbersome things like "the function which is determined by the rule $f(x) = x^2$," etc. In fact, we sometimes simply say "the function x^2," etc. The important thing to notice about all such notation is that the letter x doesn't represent any particular number. We could just as well say $f(a) = a^2$, for all numbers a; or $f(t) = t^2$, for all numbers t. The letter x or a or t just serves as a "place-holder," giving us a sensible way of saying

$$f(\text{a number}) = \text{square of that number.}$$

We could even say $f(y) = y^2$, though we usually find it convenient to reserve y for the value $f(x)$. In short, the use of any particular letter is just a convention. If we are going to call the horizontal axis the x-axis, and give numbers names like x_0, x_1, etc., then we'll probably use notation like $f(x)$; if we're thinking of a t-axis, perhaps to represent time, then we'll probably use notation like $f(t)$.

So far, we've considered only a few functions like $f(x) = cx^2$, $f(x) = x^3$, etc., and already found that there is considerable variety. Things really start to get interesting when we consider arbitrary sums, like

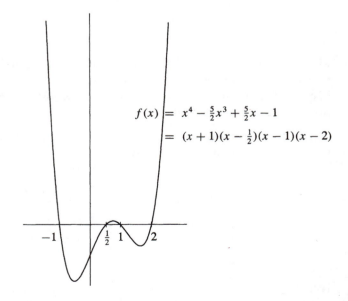

$$f(x) = x^4 - \tfrac{5}{2}x^3 + \tfrac{5}{2}x - 1$$
$$= (x + 1)(x - \tfrac{1}{2})(x - 1)(x - 2)$$

And that's just the beginning. All sorts of new complications arise if we start *dividing*. We get a graph that looks quite different from any of the previous ones when we consider a simple quotient:

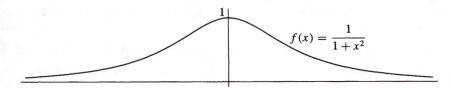

In this example we didn't have to worry about dividing by 0, since we always have $1 + x^2 > 0$. The simple example $f(x) = 1/x$, which is defined only for $x \neq 0$, shows what can happen when our numerator can have the value 0.

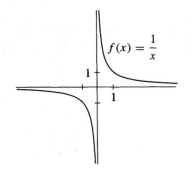

Similarly, the graph of

$$f(x) = x^2 + \frac{1}{x}$$

looks like this (where the graph of x^2 appears as a dashed line):

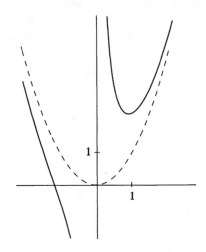

Although the functions that we have already encountered exhibit so many new features, they barely begin to represent the diversity of functions once we allow more interesting things, like functions derived from trigonometry.

Recall that for a right triangle $\triangle OCB$ with $\theta = \angle BOC$, the **sine** and

the **cosine** of the angle θ are defined by

$$\sin \theta = \frac{BC}{OB}, \qquad \cos \theta = \frac{OC}{OB}.$$

These definitions don't depend on the triangle we choose, because if the triangle $\triangle OC'B'$ also has angle θ at O, then it is similar to $\triangle OCB$, and

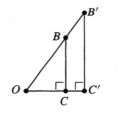

consequently

$$\frac{B'C'}{OB'} = \frac{BC}{OB}, \qquad \frac{OC'}{OB'} = \frac{OC}{OB}.$$

It is convenient to choose our triangle so that the vertex O coincides with the origin O of our coordinate system, with OC lying along the x-axis. Moreover, since the size of the triangle is irrelevant, we can also determine B by requiring it to lie on the unit circle, so that the distance

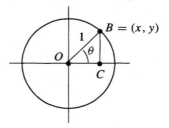

$BO = 1$, and then we simply have $\sin \theta = BC$ and $\cos \theta = OC$. Even more concisely, if $B = (x, y)$, then

$$\sin \theta = y, \qquad \cos \theta = x.$$

If we measure angles in degrees, then the angle θ in the picture above satisfies $0 < \theta < 90°$, but we can define $\sin\theta$ for all other values of θ, both positive and negative, by a scheme with which you should be familiar.

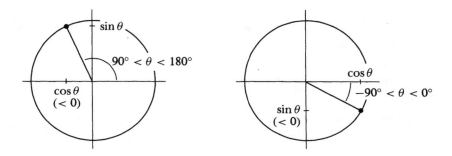

Thus we now have a function sin ("the sine function"), whose value at θ is $\sin\theta$, as well as a function cos ("the cosine function"). The graph of the sine function looks like

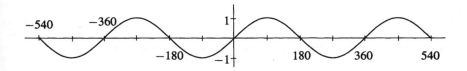

Notice that in this figure we used different scales on the two axes; without this distortion there wouldn't be much to see in our graph unless we made it extremely wide.

Aside from this aesthetic defect, measuring angles in degrees seems rather arbitrary in any case—after all, what's so special about 90, or 360? A conceptually much more natural method of measuring angles is to use the length of the arc of the unit circle from the point $A = (1, 0)$ to B.

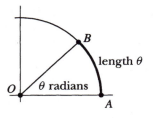

If this arc has length θ, then $\angle AOB$ is an angle of θ **radians**, or has a "radian measure" of θ.

An angle all the way around (360°) has a radian measure of 2π, the circumference of the unit circle; a right angle has a radian measure of

$$\frac{1}{4} \cdot 2\pi = \frac{\pi}{2};$$

a 45° angle thus has a radian measure of $\pi/4$, etc. Otherwise said, an angle of θ radians is

$$\frac{\theta}{2\pi} \text{ of the angle all the way around.}$$

Although we defined radian measure in terms of the length of the arc of the unit circle from $A = (1, 0)$ to B, we can just as easily specify it in

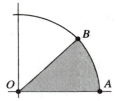

terms of the area of the sector OAB: Since an angle of θ radians is

$$\frac{\theta}{2\pi} \text{ of the angle all the way around,}$$

the area of OAB is given by

$$\text{area } OAB = \frac{\theta}{2\pi} \cdot \text{area of unit circle} = \frac{\theta}{2\pi} \cdot \pi$$
$$= \frac{\theta}{2}.$$

Thus, $\angle AOB$ is an angle of θ radians when the sector OAB has area $\theta/2$ (which conveniently eliminates mention of π).

If we now let $\sin \theta$ be the sine of an angle of θ *radians*, then the graph of sin looks like

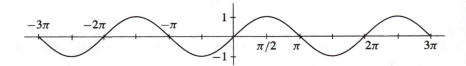

while the graph of cos looks like

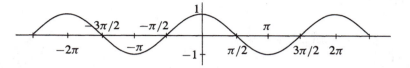

From now on, we can basically forget about degree measurement, at least so far as functions and their graphs are concerned. When mathematicians refer to the function sin, they *always* mean the function you get by using radian measure.

There's a whole contingent of other trigonometric functions, like tan, cot, etc., but they are not of primary importance, since they can be expressed in terms of sin and cos,

$$\tan \theta = \frac{\sin \theta}{\cos \theta}, \quad \text{etc.}$$

As a matter of fact, we can even express cos in terms of sin. For we have

$$(OC)^2 + (CB)^2 = 1,$$

i.e.,

$$(\cos \theta)^2 + (\sin \theta)^2 = 1,$$

which is usually written simply as

$$\cos^2 \theta + \sin^2 \theta = 1,$$

and hence

$$\cos \theta = \pm \sqrt{1 - \sin^2 \theta}.$$

The proper choice of the $+$ or $-$ sign depends on the value of θ. For

example, for $0 < \theta < \pi/2$ (between 0 and a right angle) we must choose the $+$ sign; for $\pi/2 < \theta < \pi$ (between a right angle and a straight angle) we must choose the $-$ sign, since cos is negative, while sin is still positive; etc. Because of this extra complication involving the determination of the sign, it's usually convenient to regard cos and sin as equally fundamental.

   Once we've added sin and cos to our repertoire, all sorts of new functions become available.

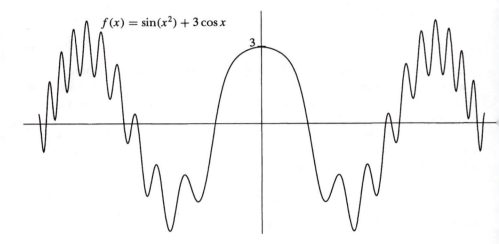

$$f(x) = \sin(x^2) + 3\cos x$$

   And there are still further ways to manufacture functions. For example, we can consider the graphs of functions like

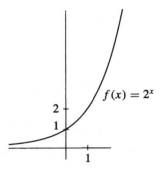

$$f(x) = 2^x$$

and

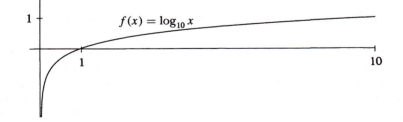

$$f(x) = \log_{10} x$$

and then all sorts of new combinations like

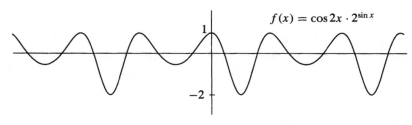

$$f(x) = \cos 2x \cdot 2^{\sin x}$$

and

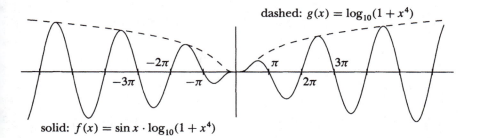

dashed: $g(x) = \log_{10}(1 + x^4)$

solid: $f(x) = \sin x \cdot \log_{10}(1 + x^4)$

But we've already seen enough to drive home the main point—there are lots and lots of functions.

We've arrived at a new world, full of wondrous things, but at the same time possibly a bit scary. While analytic geometry might be content with analyzing straight lines and circles, and perhaps a few other special curves, calculus undertakes to investigate functions in general. But this may seem almost futile. If functions are so multifarious, and exhibit such diverse features, how can we hope to say anything interesting about them all?

4 *We explore the new terrain more closely.*

As a first attempt to find some unifying theme in the vast wilderness into which we have stumbled, we're going to select one of the simplest non-linear functions, $f(x) = x^2$. Even this simple graph seems much

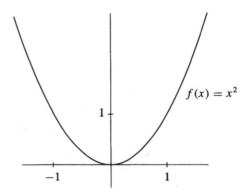

more complicated than a straight line, or even a circle, since it appears to curve by different amounts at different points.

Let's focus our attention on a particular point on the graph, say the point $P = (1, 1)$, and look at the graph more closely near this point. We're going to "look at the graph more closely" in the most literal way

possible, by taking a magnifying glass, placing it over the point P, and peering at the magnified image that we obtain.

Let's start with a magnifying glass that has a magnifying power of 5, so that it makes everything look 5 times as big as usual. Placing our magnifying glass over the point $P = (1, 1)$, we obtain the following image.

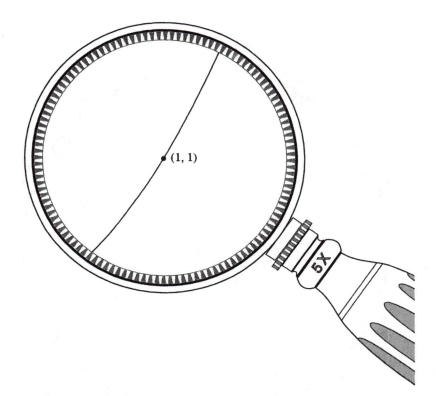

Both the original graph of $f(x) = x^2$ and the portion of the graph that appears in this figure were generated by a computer program, so we can be pretty confident that they are accurate. But our ordinary experience should immediately convince us of one of the most striking differences between the two figures:

The magnified graph curves *less* than the unmagnified one.

This is readily appreciated by voyagers on spaceships, who on earth first see the moon as a small disk in the sky, with a very obviously curved outline. As their spaceship approaches the moon, a very small portion of this outline fills up nearly the whole of the viewing port, appearing as a very small portion of a very large circle, with very gentle curvature. Even photographs from orbiting space stations illustrate this phenomenon.

Now let's suppose that we have a really high-powered magnifying glass, one that magnifies images by a factor of 100 (which we probably can't really expect to get unless we graduate from magnifying glasses to microscopes). Then we will get an image like this:

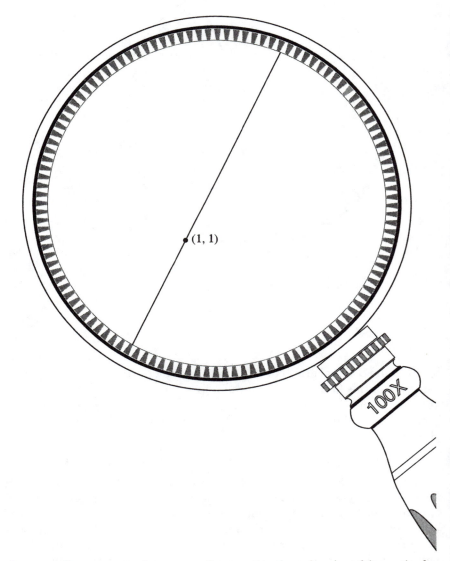

It was really pretty much a waste of computer time drawing this particular portion of the graph, because it's essentially just a straight line. Of course, it isn't really a straight line—no small portion of our graph $f(x) = x^2$ is actually a tiny straight line segment—but the amount of curvature is so small that the thickness of the ink line is far greater than the amount by which the actual image deviates from a straight line. And even if we could print a very thin line, on extremely smooth paper, with ink that had almost no tendency to smear, our eye wouldn't be able to distinguish the resulting image from a straight line.

At this point, pessimists (for whom the glass is always half-empty) may feel that we've done a lot of work for nothing. We've looked at $f(x) = x^2$ very closely, but rather than finding anything interesting, instead we've simply ended up with an unexciting straight line. On the other hand, optimists, for whom the glass is always half-full, may rejoice in this very fact. Although the graph of $f(x) = x^2$ seems to exhibit all sorts of complicated behavior, curving differently at different points, if we simply look at it closely enough, at high enough magnification, it appears to be a straight line. This is encouraging, because straight lines are the one kind of curve that we already know how to deal with.

In fact, in Chapter 2 we mentioned several different ways of writing formulas for straight lines. The *point-slope* form is obviously the one most relevant to the current situation, since, after all, we already *know* a point through which our straight line passes, namely the point $P = (1, 1)$ that we have been examining all along. That leaves just one thing that we *don't* know, namely the slope m of this straight line.

If you believe in the accuracy of the picture on page 26, then you should be able to get a good approximation to m (simply by selecting two points on the line, drawing the right triangle that they determine, and finding the ratio of the lengths of the sides of that triangle). But that doesn't

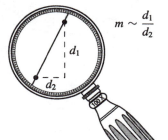

$$m \sim \frac{d_1}{d_2}$$

tell us how to calculate m without actually resorting to pictures. And this whole thing's somewhat fishy anyway, since the picture on page 26 isn't *really* a straight line in the first place!

It turns out that we'll be able to tackle these two difficulties at once, but that's a problem that we'll leave for the next chapter. At the moment we just want to emphasize that this number m *depends on the particular point P* that we happened to choose in the first place.

For example, suppose that we choose the point $P = (2, 4)$ instead. When we place the magnifying glass over this point, the image will look like

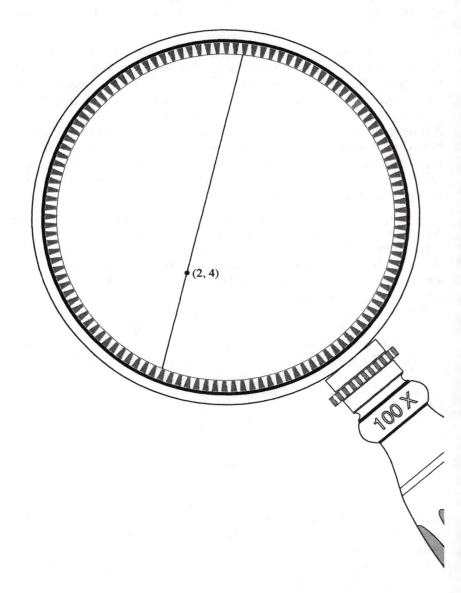

Naturally this is a completely different straight line. For one thing, it goes through the point (2, 4), rather than the point (1, 1). In addition, it has a different slope (which is again apparent if you trust in the accuracy of the pictures).

Thus, for each point of our graph we have a certain number, the slope of the straight line that the graph almost becomes when we look at a magnified portion near that point. Equivalently, given any number x_0, we can look at a magnified portion of our curve $f(x) = x^2$ near the point (x_0, x_0^2), and thereby determine some number. We might write this number as $g(x_0)$, to indicate that it depends on the original number x_0. In other words, our original function f gives us, by this process, a new function g.

It is customary to write this new function g as f' ("f prime"), and this function f' is called the **derivative** of f. Thus, $f'(x_0)$ is the slope of the straight line we obtain when we look at a magnified portion of the graph of f near the point $(x_0, f(x_0))$.

We can also simply write $f'(x)$, for the slope near the point $(x, f(x))$—while we often use x as a "place-holder" in a formula like $f(x) = x^2$, we also use x simply to indicate some arbitrary number; in practice, this won't cause confusion, since it'll be clear how we are using x.

Notice that the notation $f'(x)$ doesn't explicitly mention the second coordinate $f(x)$ of the point in question; that would be superfluous, since this second coordinate is determined once f and x are specified.

Now that we've made these definitions, all that remains is to give some substance to this whole procedure, by figuring out a reasonable way to calculate $f'(x)$ without actually resorting to magnifying glasses and rulers. The whole purpose of the next few chapters is to find out how f' can be computed, and to see why it is important.

Wandering Off On Your Own

1. Instead of the point $(1, 1)$ or $(2, 4)$ on the graph of $f(x) = x^2$, suppose that we choose the point $(0, 0)$. Then when we magnify the graph of $f(x) = x^2$ near this point we are going to get a straight line through $(0, 0)$, which will have the equation $y = mx$ for some m. The question is, what is m?

   (a) Even though we haven't figured out how to calculate m in general, you should be able to figure this out: when we magnify the graph of $f(x) = x^2$ near $(0, 0)$, there is really only one straight line through $(0, 0)$ that this magnified picture can look like.

   (b) Suppose instead that we consider the graph of $f(x) = x^4$, and once again choose the point $P = (0, 0)$. When we magnify the graph it will look essentially like some straight line through $(0, 0)$, with an equation of the form $y = mx$, and the problem is to determine m in this case: once again, there is really only one straight line that is a reasonable candidate.

5

We resort to a few calculations to secure our position.

N ow we want to get down to the business of actually calculating the slope m of the straight line we obtain when we magnify the graph

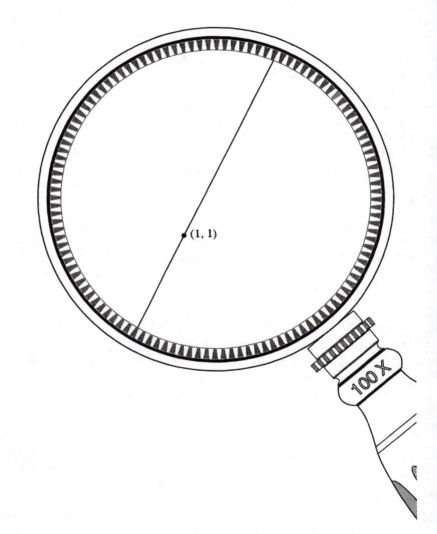

(1, 1)

100 X

of $f(x) = x^2$ near $P = (1, 1)$, one of our big problems being that this magnified graph actually isn't a straight line at all.

First we'll construct a real straight line, simply by choosing another point Q on the graph, and drawing the straight line between P and Q.

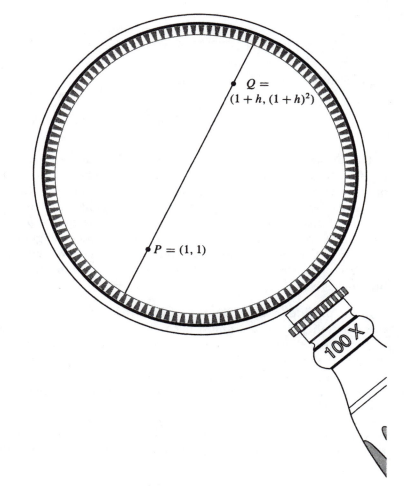

$Q = (1 + h, (1 + h)^2)$

$P = (1, 1)$

Naturally you can't see two different lines in our picture, because the original curve is so close to the new straight line through P and Q, so you just have to remember that they are really different.

Notice that we described the coordinates of Q as $(1 + h, (1 + h)^2)$. We did this as a reminder that the x-coordinate of our new point is close to the x-coordinate of the first point: Remember that our figure shows a greatly enlarged portion of a very small part of the graph of $f(x) = x^2$; consequently, *the point Q is very close to P,* which means that *h is very small.* (The letter h isn't supposed to stand for anything in particular, it's just a convenient letter that isn't used too often for anything else.)

Now that we have a specific straight line, the one through $P = (1, 1)$ and $Q = (1 + h, (1 + h)^2)$, it is easy to write down the slope of this straight line: it is

$$(1) \qquad m = \frac{(1 + h)^2 - 1}{(1 + h) - 1} = \frac{(1 + h)^2 - 1}{h}.$$

This can be simplified even further, as

$$(2) \qquad m = \frac{1 + 2h + h^2 - 1}{h} = \frac{2h + h^2}{h} = 2 + h.$$

This answer doesn't immediately give us the number m we are looking for. In fact, it doesn't give us a specific number at all—it depends on h, which is hardly surprising, since it depends on our choice of the point Q. But as we consider larger and larger magnifications, we will have to choose Q closer and closer to P, and thus choose h closer and closer to 0, so the slopes, $2 + h$, of these straight lines will get closer and closer to 2. Consequently, although our magnified curve doesn't actually ever become a straight line, it becomes closer and closer to the straight line through $P = (1, 1)$ that has the slope $m = 2$.

In the previous chapter we spoke of the straight line that the graph of the function $f(x) = x^2$ "almost becomes" when we look at a magnified portion near the point $P = (1, 1)$. Our simple calculations have now made that embarrassing phrase much clearer: If we look at actual straight lines from P to nearby points, then these lines come closer and closer to (that is, their slopes become closer and closer to the slope of) the line through P with slope equal to 2.

As we indicated at the end of the previous chapter, this slope is usually denoted by $f'(1)$, so that we can summarize our calculations in the short equation

$$f'(1) = 2.$$

(Remember that in the notation $f'(1)$, the 1 is the first coordinate of the point we are interested in, the second coordinate being $f(1)$; for $f(x) = x^2$, with $f(1) = 1$, we are looking at the point $(1, 1)$.)

The fact that the slopes in equation (1) approach closer and closer to 2 as we make h closer and closer to 0 is usually written

$$(3) \qquad \lim_{h \to 0} \frac{(1 + h)^2 - 1}{h} = 2,$$

and in general, $\lim_{h \to 0} g(h)$ denotes the number that $g(h)$ gets close to as we choose h closer and closer to 0. Thus the equation

$$\lim_{h \to 0} g(h) = A,$$

which is read "the limit of $g(h)$ as h approaches 0 equals A," simply means that $g(h)$ approaches closer and closer to A as we choose h closer and closer to 0.

To sum up, we are *defining* $f'(1)$ to mean

$$f'(1) = \lim_{h \to 0} \frac{(1+h)^2 - 1}{h},$$

and we have *calculated* that this limit has the value 2.

If we instead look at the point $P = (2, 4)$ on the graph of $f(x) = x^2$ and choose a point Q close to P, of the form $(2 + h, (2 + h)^2)$ for small h,

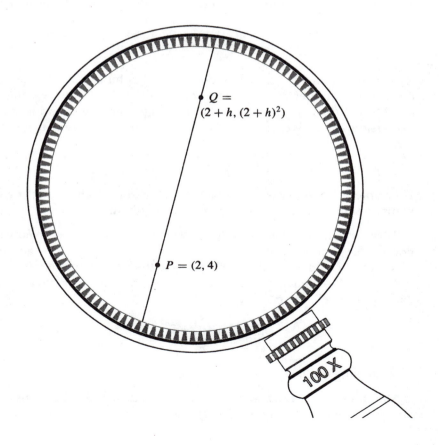

then the slope of the straight line through P and Q is

(4) $$m = \frac{(2+h)^2 - 2^2}{(2+h) - 2} = \frac{(4 + 4h + h^2) - 4}{h} = \frac{4h + h^2}{h} = 4 + h.$$

As we choose h closer and closer to 0 (or equivalently, as we choose Q closer and closer to P), these slopes approach 4, in symbols,

(5) $$\lim_{h \to 0} \frac{(2+h)^2 - 2^2}{h} = 4,$$

which means that our magnified curve becomes closer and closer to the straight line through $(2, 4)$ with slope 4. Thus

$$f'(2) = 4.$$

It should be emphasized that an equation like (5) only means that

$$\frac{(2+h)^2 - 2^2}{h}$$

is close to 4 when h is *close* to 0; it doesn't say anything about the value of this expression when $h = 0$. In fact, for $h = 0$ we obtain the meaningless expression $0/0$. The only reason that we can predict the behavior of this expression when h is close to 0 is because we were able to rewrite it in the form

$$\frac{(2+h)^2 - 2^2}{h} = 4 + h.$$

In Chapter 9 we will see an example where a simple algebraic trick of this nature doesn't work, so that more work is required.

By the way, notice that the letter h (which we chose rather arbitrarily) doesn't have any special significance in an equation like (5), and any other letter could be used as well. Just like the letter x in the definition "$f(x) = x^2$," the letter h in (5) is just a "place-holder," and equation (5) could just as well be written as

$$\lim_{u \to 0} \frac{(2+u)^2 - 2^2}{u} = 4,$$

or something similar.

After these initial calculations, we might as well ask what happens at an arbitrary point on the graph of $f(x) = x^2$, which we can call $P = (a, a^2)$.

Again we take a point $Q = (a + h, (a + h)^2)$ close to P and compute that the slope of the straight line through P and Q is

(6) $$\frac{(a + h)^2 - a^2}{(a + h) - a} = \frac{a^2 + 2ah + h^2 - a^2}{h} = \frac{2ah + h^2}{h} = 2a + h.$$

Since we clearly have

$$\lim_{h \to 0} 2a + h = 2a,$$

this means that the straight line we are seeking, through $P = (a, a^2)$, has slope $2a$. Thus we have the general formula

$$f'(a) = 2a.$$

By the way, we can just as well write this as

$$f'(x) = 2x;$$

we simply used a to emphasize that we were looking at a fixed point (a, a^2) on the graph of $f(x) = x^2$. Summing up,

$$\boxed{\text{If } f(x) = x^2, \text{ then } f'(x) = 2x.}$$

Remember, once again, that we are *defining*

$$f'(x) = \lim_{h \to 0} \frac{f(x + h) - f(x)}{h},$$

and we have *calculated* that $f'(x) = 2x$. This point is even more important than all the work that was involved in this "differentiation," as the calculation of derivatives is usually called. It is the notion of a *limit* that allows us to get around the fact that no matter how much we magnify a graph it doesn't really become a straight line: instead of worrying about that, we simply compute the slope of the straight lines through a given point P and nearby points Q and then see what number these slopes are getting close to as we choose Q closer and closer to P.

To get a somewhat better idea of what sort of calculations differentiation might entail, let's turn out attention to the slightly more complicated function $f(x) = x^3$.

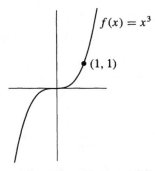

Let us again consider the point $P = (1, 1)$, which also happens to be on the graph of this new function f. Taking a point $Q = (1+h, (1+h)^3)$

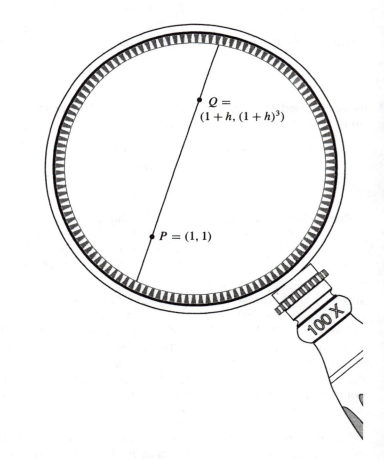

near P we compute that the slope of the straight line passing through the points P and Q is

$$\frac{(1+h)^3 - 1}{(1+h) - 1} = \frac{(1 + 3h + 3h^2 + h^3) - 1}{h}$$
$$= \frac{3h + 3h^2 + h^3}{h}$$
$$= 3 + 3h + h^2.$$

Although this expression is a little more complicated than the expression $2 + h$ that appears in (2), it is just as easy to see what happens when h gets smaller and smaller: the product $3h$ gets closer and closer to 0, and the same is true of h^2 (if h is small, then so is h^2, in fact, h^2 is usually even closer to 0 than h!). So the sum also gets closer and closer to 0, which means that

$$\lim_{h \to 0} 3 + 3h + h^2 = 3.$$

So the straight line we are seeking, through the point $P = (1, 1)$, has slope 3.

In other words, for $f(x) = x^3$, we have

$$f'(1) = \lim_{h \to 0} \frac{f(1+h) - f(1)}{h} = \lim_{h \to 0} 3 + 3h + h^2 = 3.$$

Rather than choosing another particular point on the graph, we'll immediately consider a general point $P = (a, a^3)$ on the graph of $f(x) = x^3$. We take a nearby point $Q = (a + h, (a + h)^3)$, and compute that the slope of the straight line through P and Q is

$$\frac{(a+h)^3 - a^3}{(a+h) - h} = \frac{a^3 + 3a^2h + 3ah^2 + h^3 - a^3}{h}$$
$$= \frac{3a^2h + 3ah^2 + h^3}{h}$$
$$= 3a^2 + 3ah + h^2.$$

Thus for $f(x) = x^3$ we have

$$f'(a) = \lim_{h \to 0} \frac{(a+h)^3 - a^3}{(a+h) - h}$$
$$= \lim_{h \to 0} 3a^2 + 3ah + h^2$$
$$= 3a^2.$$

Once again, we can also write this as $f'(x) = 3x^2$, so that we have the result

$$\boxed{\text{If } f(x) = x^3, \text{ then } f'(x) = 3x^2.}$$

We should mention that the notation $f'(x)$ is used interchangeably with an older notation for the derivative,

$$f'(x) = \frac{df(x)}{dx}$$

(read "the derivative of $f(x)$ with respect to x"), or even

$$f'(x) = \frac{d}{dx}f(x).$$

It ought to be fairly obvious that in such notation the symbol d does not stand for a number (which could simply be canceled in the numerator and denominator)—it is really the whole "differentiation" symbol

$$\frac{d}{dx}$$

that has a meaning, namely "take the derivative of."

The reason for such strange notation is that in the definition

$$f'(x) = \lim_{h \to 0} \frac{f(x+h) - f(x)}{h}$$

the symbol

$$h \text{ is sometimes replaced by } \Delta x,$$

using the Greek letter **Delta** for the "**D**ifference in x" (i.e., the difference between the x-coordinate of our fixed point P and our other point Q), and similarly we may write

$$f(x+h) - f(x) = \Delta f,$$

where Δf now represents the difference in the values of f at P and Q. With these abbreviations, we can then write

$$f'(x) = \lim_{\Delta x \to 0} \frac{\Delta f}{\Delta x}.$$

The $\dfrac{d}{dx}$ notation is meant as a sort of symbolic abbreviation for this limit: replacing Δ by d is meant to indicate that we are taking the limit as $\Delta \to 0$.

Although this "d-notation" may seem a lot more complicated than f', in many cases it actually works out to be much briefer. For example,

$$\frac{dx^2}{dx} \text{ means: } f'(x), \text{ when } f(x) = x^2$$

$$\frac{dx^3}{dx} \text{ means: } f'(x), \text{ when } f(x) = x^3$$

etc.,

so one can write

$$\frac{dx^2}{dx} = 2x,$$

$$\frac{dx^3}{dx} = 3x^2,$$

without ever having to name functions. The reason for this economy of notation is obviously the fact that the $\dfrac{d}{dx}$ notation uses x as a "place-holder" in the very same way that x functions as a "place-holder" when we define f by saying "$f(x) = x^2$." Thus one could just as well write

$$\frac{dt^2}{dt} = 2t,$$

$$\frac{dt^3}{dt} = 3t^2,$$

and these equations would mean precisely the same as the previous ones.

The one place where the $\dfrac{df(x)}{dx}$ notation becomes more complicated is when we need to indicate the derivative at a specific point, like a. In this case we use something like

$$\left. \frac{df(x)}{dx} \right|_{x=a},$$

which indicates that we first find $f'(x)$, as a formula involving x, and then set $x = a$ in this formula to get $f'(a)$.

As a final illustration of a differentiation, we'll consider

$$f(x) = \frac{1}{x}$$

(which is only defined for $x \neq 0$). Now we have

$$f'(x) = \lim_{h \to 0} \frac{f(x+h) - f(x)}{h}$$

$$= \lim_{h \to 0} \frac{\dfrac{1}{x+h} - \dfrac{1}{x}}{h}$$

$$= \lim_{h \to 0} \frac{\dfrac{x - (x+h)}{x(x+h)}}{h}$$

$$= \lim_{h \to 0} \frac{-h}{x(x+h)h}$$

$$= \lim_{h \to 0} \frac{-1}{x(x+h)}$$

$$= -\frac{1}{x^2}.$$

We can write this as

$$\frac{dx^{-1}}{dx} = -\frac{1}{x^2} = -1x^{-2},$$

which makes it fit into a pattern with

$$\frac{dx^2}{dx} = 2x,$$

$$\frac{dx^3}{dx} = 3x^2.$$

In fact, in general we have

$$\frac{dx^n}{dx} = nx^{n-1}.$$

But the details of such calculations will be left to your calculus course, where you will also learn important differentiation rules for functions that are built up out of other functions.

6   *A brief interlude,*
just to provide continuity.

I n the previous chapter we encountered the notion of a limit, together with the notation

$$\lim_{h \to 0} g(h) = A.$$

Recall that the letter h is just a "place-holder," so we can equally as well write

$$\lim_{x \to 0} g(x) = A.$$

Even more generally, we write

$$\lim_{x \to a} g(x) = A$$

("the limit of $g(x)$ as x approaches a equals A") to express the fact that $g(x)$ approaches closer and closer to A as we choose x closer and closer to a.

In your calculus course you will probably encounter some simple rules about limits, like

$$\lim_{x \to a} f(x) + g(x) = \lim_{x \to a} f(x) + \lim_{x \to a} g(x),$$
$$\lim_{x \to a} f(x) \cdot g(x) = \lim_{x \to a} f(x) \cdot \lim_{x \to a} g(x),$$

which are used in proving the rules for derivatives mentioned at the end of the last chapter.

There is also the rule

$$\lim_{x \to a} \frac{f(x)}{g(x)} = \frac{\lim_{x \to a} f(x)}{\lim_{x \to a} g(x)},$$

which holds *provided that* $\lim_{x \to a} g(x) \neq 0$. This extra condition is hardly a minor matter, for all the limits we encountered in calculating derivatives were precisely of the form

$$\lim_{h \to 0} \frac{A(h)}{B(h)},$$

where both $A(0)$ and $B(0)$ are 0; thus we couldn't simply plug in the value $h = 0$, and instead had to rely on algebraic trickery.

41

Exactly the same problem occurs with a limit like

$$\lim_{x \to a} \frac{x^2 - a^2}{x - a},$$

where again we cannot simply plug in the value $x = a$, but have to use some sort of algebraic simplification. By the way, notice that this expression is really just another way of writing

$$\lim_{h \to 0} \frac{(a + h)^2 - a^2}{(a + h) - a}$$

—we are simply substituting $x - a$ for h in this expression.

But we could also consider simpler things like

$$\lim_{x \to a} x^2, \qquad \lim_{x \to a} x^3, \qquad \text{etc.,}$$

which obviously have the values

$$a^2, \qquad\qquad a^3, \qquad\qquad \text{etc.}$$

In fact, one might be tempted to say that we always have

$$\lim_{x \to a} g(x) = g(a),$$

so long as $g(a)$ really makes sense.

But this can easily fail to be true for certain functions that arise quite naturally. For example, the function

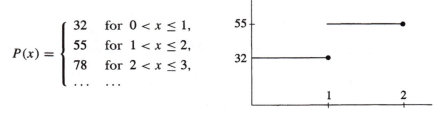

$$P(x) = \begin{cases} 32 & \text{for } 0 < x \leq 1, \\ 55 & \text{for } 1 < x \leq 2, \\ 78 & \text{for } 2 < x \leq 3, \\ \dots & \dots \end{cases}$$

is used by the Post Office to determine the postage of a letter in terms of its weight in ounces (at least that function was used when this book was written; who knows what it'll be by the time you read it). Here

it is definitely **false** that $\lim_{x \to 1} P(x) = P(1) = 32$:

It is true that $P(x)$ will be close to 32 (in fact, equal to 32) for those x close to 1 that are smaller than 1; but when we choose x close to 1 that are greater than 1, the value of $P(x)$ will not be close to 32 (it will be 55). In fact, in this case $\lim_{x \to 1} P(x)$ doesn't even make sense, or as we usually say, "the limit does not exist."

Functions of this sort invite all kinds of philosophical questions that the Post Office might not want to take too seriously (can an extra ink drop on a letter really change the postage?),[1] but whether or not the function P makes much sense in the real world, it has to be accounted for in the world of mathematics. So we want to adopt some terminology to distinguish "reasonable" functions from such oddities. We say that a function f is **continuous at** a if

$$\lim_{x \to a} f(x) = f(a).$$

Thus the function P is *not* continuous at 1.

The first function shown below is a modification of P that is clearly also not continuous at 1, and the accompanying functions are also clearly

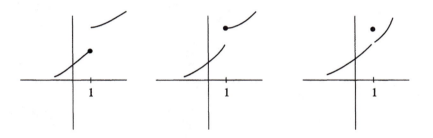

not continuous at 1 [in the third case $\lim_{x \to 1} f(x)$ exists, but it is $\neq f(1)$]. Actually, there can be all sorts of other horrible ways for a function not

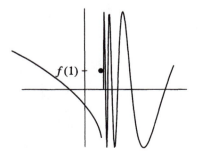

to be continuous at a, but most of the time we are going to be ignoring

---

[1] Electronic mail doesn't have these problems—everyone realizes that an extra byte might very well end up costing a bit more.

all such problems. We will usually be considering only those functions f that satisfy the condition

f is continuous at a for all a;

for brevity, such functions are simply called **continuous**.

There are many situations where non-continuous functions can be considered in calculus, but the most important aspects of calculus can all be illustrated and understood in terms of continuous functions alone. In fact, from the point of view of calculus even continuous functions can be unreasonable.

For example, the function $f(x) = |x|$ (page 13) is continuous, but if we place our magnifying glass over the point $(0, 0)$ on the graph, the

magnified image looks exactly the same as the original—so in this case it never looks anything like a straight line! Indeed, if we try to calculate

$$f'(0) = \lim_{h \to 0} \frac{f(h) - f(0)}{h} = \lim_{h \to 0} \frac{f(h)}{h}$$

$$= \lim_{h \to 0} \begin{cases} \dfrac{h}{h} & h > 0 \\[2mm] \dfrac{-h}{h} & h < 0 \end{cases}$$

$$= \lim_{h \to 0} \begin{cases} 1 & h > 0 \\ -1 & h < 0, \end{cases}$$

we immediately see that the limit doesn't exist. Generally, we say that f is **differentiable** at a if $f'(a)$ exists (i.e., if the limit used to define $f'(a)$ exists); if f is not differentiable at a, then the symbol $f'(a)$ doesn't even make sense.

7

*We settle down for a rest, and the author
goes off on a tangent, reminiscing about his youth.*

Unlike the majority of my fellow classmates in high school, I enjoyed most of my math classes, and I particularly liked geometry (well, ... , mathematicians are funny people, you already knew that). It especially impressed me when an intuitive idea could be given a precise definition. For example, I was quite pleasantly surprised when a right angle ∠*ABC* was simply defined to be one that equals its complementary

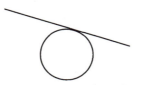

angle ∠*DBC*—this short definition obviously expressed precisely the idea that a perpendicular line doesn't fall more to one side than the other.

However, I wasn't quite as comfortable later on when a **tangent line** to a circle was defined as a straight line that intersects the circle only once.

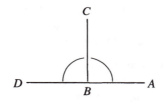

I probably wouldn't have been able to put my unease into words at the time, but there's clearly something unsatisfactory about this definition. For example, any one can see that the straight line in the figure below is

tangent to the circle, even though only a very small portion of the circle has been drawn, so that we are in no position to check that the line doesn't intersect the circle somewhere else. Similarly, the straight line in the next figure clearly *isn't* tangent to the circle, even though the other point of

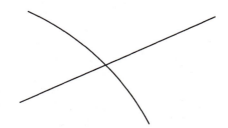

intersection isn't in the picture, and in fact, is somewhere off the page entirely. This criterion, of intersecting the circle only once, doesn't seem to capture the intuitive idea that the tangent line just "grazes" the circle.

To be sure, even if the definition doesn't seem to express our intuitive idea, it is correct, and one can use it to prove all the results one expects in plane geometry. But our objections become at lot more substantial when we consider curves other than circles.

For example, what about a parabola? The straight line in this picture,

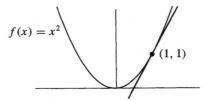

which certainly seems to be the one that we would select as the tangent line, does indeed have the property that it intersects the parabola only once. The problem is, this line is not the only one with this property: in fact, the vertical line through this point also has this property, and

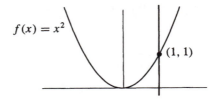

we certainly wouldn't want to call it a tangent line!

The problem becomes even more acute when we consider slightly more complicated curves, for example, the graph of $f(x) = x^3$. It would seem

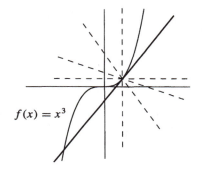

$$f(x) = x^3$$

reasonable to call the solid straight line in this picture a tangent line—it appears to "graze" the curve in the same way that the tangent line to a circle or a parabola does—but this straight line intersects the curve in more than one point. Various other straight lines that *do* intersect the curve just once (dashed lines in the figure) certainly *shouldn't* be called tangent lines!

Although we've been using the rather vague word "graze" in this discussion, we can give some substance to the concept by using the very same technique that we employed in Chapter 4. The notion that a straight line "grazes" a curve at some point P is one that involves the behavior of the curve very close to P, rather than at some distant point. So we will put our magnifying glass over this point, and see what sort of image we obtain.

Let's consider the parabola $f(x) = x^2$ again, together with the point $P = (1, 1)$ on the parabola, and the vertical straight line through this point.

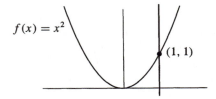

If we place a magnifying glass that magnifies everything by a factor of 5 over the point P, we get the image

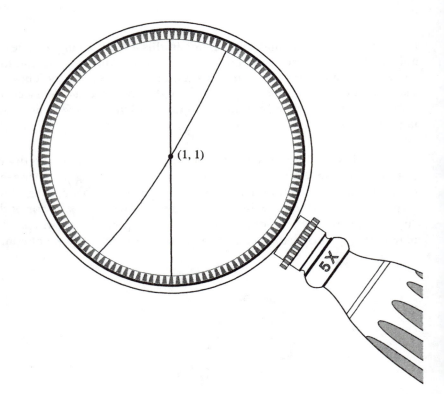

The vertical straight line stays straight, of course, while the magnified portion of the parabola shows only a small amount of curvature.

When we magnify everything by a large factor, like 100, our vertical straight line still remains the same, and our parabola becomes practically another straight line:

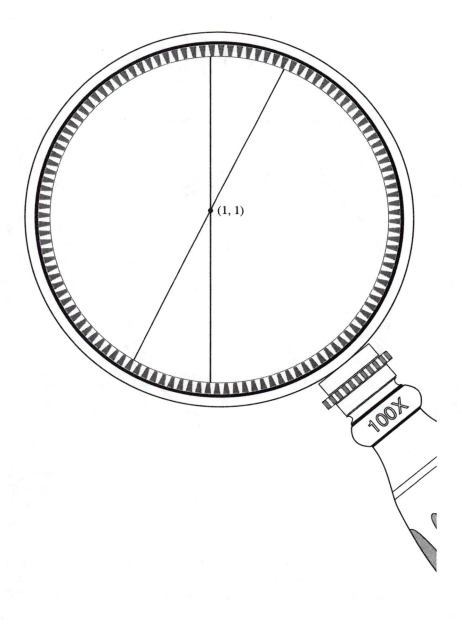

Things look quite a bit different, however, if we start with the tangent line to the parabola at $P = (1, 1)$.

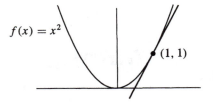

Magnifying everything by a factor of 5 produces the image

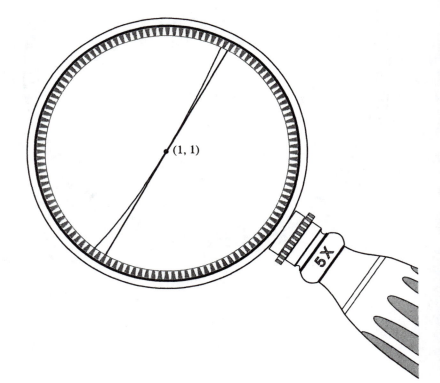

Of course, the straight tangent line remains a straight line, but because the tangent line grazes the parabola, the magnified version still grazes the magnified portion of the parabola, which again shows only a small amount of curvature.

What happens, then, when we magnify everything by a factor of 100? Our parabola becomes practically a straight line, *and this straight line coincides with the magnified image of our tangent line.* In other words, the tangent

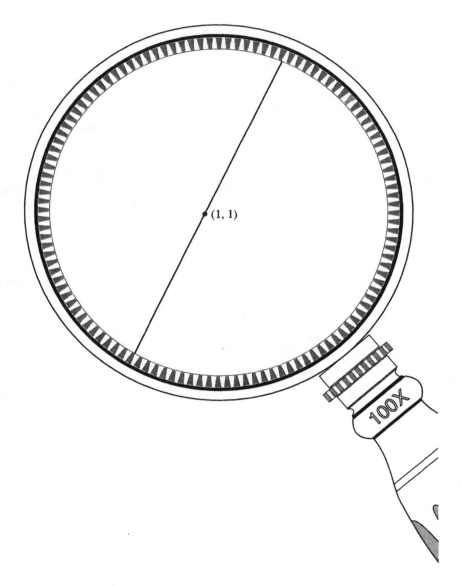

line *is* the straight line that we first discovered back in Chapter 4, when we magnified our curve by a very large amount. More precisely, we might as well simply *define* the tangent line to be this line.

Thus, to put things in a logical order, for a function f we define

$$f'(a) = \lim_{h \to 0} \frac{f(a+h) - f(a)}{h},$$

and we then *define* the tangent line to the graph of f at the point $(a, f(a))$ to be the line through this point with slope $f'(a)$. Since we've now defined a tangent line in terms of the point it goes through and its slope, we can write down an explicit equation for this straight line by using the point-slope form of Chapter 2.

That was all said in a very general way, so let's be more specific, and ask for the equation of the tangent line of the parabola $f(x) = x^2$ at the point $P = (1, 1)$. In Chapter 5 we found that the slope of this line was

$$f'(1) = 2 \cdot 1 = 2.$$

Since it goes through the point $P = (1, 1)$, this line therefore has the point-slope form

$$y - 1 = 2(x - 1),$$

or the equivalent slope-intercept form

(1) $$y = 2x - 1.$$

Now that we have the equation of this tangent line, we can easily check that it really does intersect the parabola $f(x) = x^2$ just once, just as in the picture. In fact, if a point (x, y) satisfies both

$$\begin{cases} y = x^2 & \text{so that it is on the graph of } f(x) = x^2, \\ y = 2x - 1 & \text{so that it is on the tangent line, with equation (1)} \end{cases}$$

then we have

$$x^2 = 2x - 1,$$

or

$$0 = x^2 - 2x + 1.$$

And sure enough, this equation has only the solution $x = 1$, since it can be written as

$$0 = x^2 - 2x + 1 = (x - 1)^2.$$

There's no need to fix our attention on the point $P = (1, 1)$; we can just as well consider an arbitrary point $P = (a, a^2)$ on the parabola $f(x) = x^2$. We know that

$$f'(a) = 2a,$$

so the equation of the tangent line, which goes through (a, a^2) and has slope $2a$, is

$$y - a^2 = 2a(x - a),$$

or

$$y = 2ax - a^2.$$

A Yawn And A Stretch After The Rest

1. Check that this straight line intersects the parabola only at the point (a, a^2).

2. (a) For the function $f(x) = x^3$, show that the equation of the tangent line through the point (a, a^3) is

$$y = 3a^2x - 2a^3.$$

   (b) Show that if (x, y) is on the intersection of this tangent line and the graph of $f(x) = x^3$, then

$$x^3 - 3a^2x + 2a^3 = 0.$$

   (c) This equation naturally has the root $x = a$, so $x^3 - 3a^2 + 2a^3$ is divisible by $(x - a)$. Show that

$$\begin{aligned} x^3 - 3a^2x + 2a^3 &= (x - a)(x^2 + ax - 2a^2) \\ &= (x - a)(x - a)(x + 2a). \end{aligned}$$

   (d) At what other points does the tangent line intersect the graph of $f(x) = x^3$? Make sure your answer looks reasonable in a picture.

8 *Getting back up to speed.*

P arabolas, which we've used so often as examples, arose quite naturally in Chapter 3, as the graphs of the simplest non-linear functions. But we also see them all the time in everyday life—at least if everyday life includes sports (whether experienced through active participation, or as a couch potato). Baseballs, basketballs, footballs, volleyballs, golf balls, and just about anything else that is thrown or hit, moves in a parabola, although this parabola is "upside down".

The parabolic path of these projectiles is directly related to the physical law determining the rate at which something falls under the force of gravity. For example, if a ball is thrown horizontally from a tower, then (ignoring air resistance) it maintains a constant horizontal motion, while

$$s(t) = 16t^2$$

its downward vertical distance $s(t)$ at time t increases according to the rule

$$s(t) = 16t^2,$$

where $s(t)$ is measured in feet and t in seconds. The resulting path is thus part of a parabola, with the tower as the maximum height.

We usually throw balls at an angle, so that they not only have a horizontal velocity, but also an initial upward velocity. In this case we again obtain a parabolic path, but with a greater maximum height; the part of the path before this maximum height is just the reverse of the path that would have been followed if the ball had been released at the top in the other direction.

In order to concentrate on a single function, we will simply consider the downward vertical distance $s(t) = 16t^2$ and ignore the horizontal motion. So we are basically considering a ball that is dropped from some height, rather than thrown (or perhaps a ball thrown directly upwards).

As soon as we consider something moving, it is natural to ask about its velocity. For example, when catching a ball we learn to anticipate its velocity, since this tends to impress itself quite forcibly upon us, possibly requiring the use of special gear like baseball gloves. Rather crude experiments of this sort make it clear that the velocity depends upon the height from which the ball descends; we quickly learn, even as toddlers, that the velocity is going to be greater when that height is greater.

When we have the simple formula

$$s(t) = 16t^2$$

for the distance traveled after time t, we certainly ought to be able to compute the velocity with which the ball is travelling at a particular time t_0, and thus predict how much impact to expect. The only problem is that our ordinary definition of velocity,

$$(*) \qquad \text{velocity} = \frac{\text{distance traveled}}{\text{time}},$$

is really a definition of *average velocity*. If something is moving with "uniform velocity" (so that the distance satisfies $s(t) = c \cdot t$ for some constant c), then this uniform velocity is a useful concept, but it becomes a lot less useful in general.

For example, when we have the formula

$$s(t) = 16t^2$$

we might try to figure out the impact to be expected at time t_0 by calculating the average velocity that the ball would have if it had been allowed to travel for another second. Thus, we would use (∗) over the time interval from t_0 to $t_0 + 1$, and get

$$\text{average velocity} = \frac{16(t_0 + 1)^2 - 16t_0{}^2}{1}$$
$$= \frac{16t_0{}^2 + 32t_0 + 1 - 16t_0{}^2}{1}$$
$$= 32t_0 + 1.$$

On the other hand, if we calculate the average velocity that the ball would have if it had been allowed to travel for only another .1 seconds, then we would get

$$\text{average velocity} = \frac{16(t_0 + .1)^2 - 16t_0{}^2}{.1}$$
$$= \frac{16t_0{}^2 + 3.2t_0 + .01 - 16t_0{}^2}{.1}$$
$$= \frac{3.2t_0 + .01}{.1}$$
$$= 32t_0 + .1.$$

It shouldn't be very surprising that we get a smaller answer when we calculate the average velocity over the smaller time interval, of only .1 seconds after the current time t_0. Since the ball is falling faster and faster, the longer time interval includes a period during which the ball would be moving faster if we hadn't caught it at time t_0; averaging over that longer time interval thus gives too large an answer for determining the force with which the ball will hit at time t_0.

Of course, the same objection applies even to the time interval of .1 seconds: we'll get a better answer by considering an even smaller time interval, say .01 seconds:

$$\text{average velocity} = \frac{16(t_0 + .01)^2 - 16t_0{}^2}{.01}$$
$$= \frac{16t_0{}^2 + .32t_0 + .001 - 16t_0{}^2}{.01}$$
$$= \frac{.32t_0 + .001}{.01}$$
$$= 32t_0 + .01.$$

By now it should be clear that we might as well simply consider an arbitrary small time interval h and compute that at an arbitrary time, which we'll go back to calling t, the average velocity on the time interval from t to $t + h$ is

$$\text{average velocity} = \frac{16(t + h)^2 - 16t^2}{h}$$
$$= \frac{16t^2 + 32ht + h^2 - 16t^2}{h}$$
$$= \frac{32ht + h^2}{h}$$
$$= 32t + h.$$

This calculation ought to look pretty familiar—it's just a calculation for the derivative of the function $s(t) = 16t^2$—and it gives better and better answers for the velocity at time t as we choose h smaller and smaller. This clearly means that the "best" or "right" answer for the velocity ought to be $32t$, the result when we take the limit of these quotients. In other words, if we consider something moving along a line according to the rule

$$s(t) = 16t^2,$$

then we should define the velocity at time t to be the derivative, $32t$.

More generally, if the motion along a line is given by any function s, then we define the "**(instantaneous) velocity**" at time t to be

$$s'(t) = \lim_{h \to 0} \frac{s(t + h) - s(t)}{h}.$$

Notice that we haven't proved a theorem, since until now the notion of **velocity** (as opposed to *average velocity*) simply wasn't defined: but we have demonstrated that this definition is the most reasonable one! Thus we see that derivatives (and limits, which are used to define them) have a natural place in physics. In fact, the derivative original arose in physics for precisely this purpose.

Note also that with our $\frac{d}{dt}$ notation we would write

$$\text{velocity} = s'(t) = \frac{ds}{dt},$$

which explains why we denoted the distance travelled by s [which stands for the Latin *situs* = position] rather than by d.

There are only two small clarifications that need to be made concerning the previous discussion.

First of all, with our definition, **velocity** can be positive or negative. If we measured distances from the earth's surface upwards, so that our falling object satisfies an equation of the form

$$s(t) = H - 16t^2$$

for some initial height H, then

$$s'(t) = -32t,$$

which merely says that the falling object is moving downwards. Sometimes **speed** is defined as the absolute value $|s'(t)|$.

Note also that our definition of velocity has now been made only for an object moving along a straight line. It thus applies to the case of a falling object, but *not* to a thrown object, which moves along a plane curve.

The general question of velocity for objects moving in a plane, or in space, is properly the subject matter of advanced calculus (or at least a little more advanced than the calculus we are considering), but we can at least mention the answer in the case of an object that has been thrown, so that its horizontal velocity is a constant c, while its downward vertical distance is $s(t)$, so that at time t the object is as the point $(c \cdot t, s(t))$.

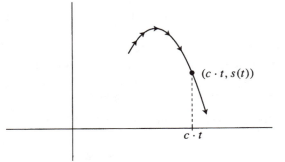

In this case, we have a "horizontal velocity," represented by a horizontal arrow of length c, and a "vertical velocity," represented by a vertical arrow of length $s'(t)$. These two arrows are the side of a rectangle, and the diagonal of this rectangle gives us a new arrow. Now this *arrow* is what we call the **velocity** in this case. So the velocity is, strictly speaking, an

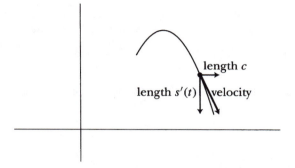

arrow, not just a number. However, we can then define the **speed** to be the length of this arrow, which in this case would be

$$\sqrt{c^2 + [s'(t)]^2}.$$

Something To Worry About

1. How many seconds do you have to get out of the way of a penny that falls from a 400 foot high building? If you don't make it, how fast will the penny be going when it hits you?

    (All our calculations have ignored air resistance. In some cases this can limit the final velocity significantly [no one worries about the height from which rain drops are falling], and even a penny will eventually be limited to a terminal velocity. A dropped penny can be lethal, however, and so can a bullet shell that was aimed into the air and falls back down empty.)

9

Emboldened by our progress thus far, we resolutely begin investigating sin (a truly exhausting endeavor!).

B y now we have discovered numerous new concepts, but, as is usual in such a discovery process, the ideas didn't necessarily arrive in the most logical order. So it might be a good idea to quickly summarize our journey so far.

1. The first idea (logically speaking) is the notion of *limit*,

$$\lim_{x \to a} f(x) = A.$$

This notion can be used to define *continuity*,

$$\lim_{x \to a} f(x) = f(a),$$

an equation which expresses the fact that the function f doesn't jump or do anything wild near the number a.

But limits were also used for a much more important purpose.

2. For the special limit

$$f'(x) = \frac{df(x)}{dx} = \lim_{h \to 0} \frac{f(x+h) - f(x)}{h},$$

which we called the *derivative*, the numerator and denominator both have the value 0 for $h = 0$, so that there is no question of evaluating the limit without at least a little algebraic trickery.

We supplied enough calculations to find $f'(x)$ for a few functions; using results about limits, and the right algebraic tricks, it is then possible to obtain formulas for many other functions.

3. The derivative turns out to have an important geometric interpretation: $f'(a)$ is the *slope* of the tangent line to the graph of f at the point $(a, f(a))$.

4. In addition, the derivative turns out to have an important physical interpretation: when $s(t)$ represents position at time t, the derivative $s'(t)$ represents *velocity* at time t.

We didn't spend much time stating theorems about limits or derivatives, although we did supply enough calculations to find $f'(x)$ for a few simple functions. In all such cases, our calculations were possible because the formula for

$$\frac{f(x+h)-f(x)}{h}$$

could be simplified algebraically in a way that made the result obvious. Now we're going to be more adventurous, and try to find the derivative of the function $f(x) = \sin x$.

Recall that the graph of the function $f(x) = \sin x$ is a curve that is quite

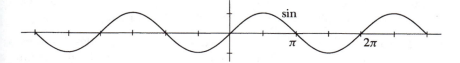

attractive, indeed quite sinuous. More precisely, it has this aesthetically pleasing form when we use **radian measure**. Recall this means that for a number h, we choose the point B on the unit circle so that the arc $\overset{\frown}{AB}$ of

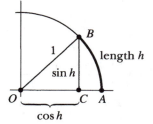

this circle has length h; then $\sin h$ is the length of the perpendicular BC, while $\cos h$ is the length of the line OC.

Recall also that instead of measuring the length of $\overset{\frown}{AB}$ we could instead measure areas: the angle AOB is an angle of h radians when the shaded

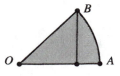

sector AOB has area given by

(A) $$\text{area } \blacktriangle = \frac{h}{2}.$$

Now we're ready to tackle the problem of finding

$$\sin'(x) = \lim_{h \to 0} \frac{\sin(x + h) - \sin x}{h}.$$

If all you care about is the answer, you can just turn to pages 66–68, and in any case you shouldn't feel that you have to be able to reproduce the following calculations (I had to peek at a book to refresh my memory). But it's worth reading through the argument once, just to get some idea of the sorts of maneuvers that may be required to find derivatives—after all, you wouldn't want to go through life with the mistaken impression that it's always just a bunch of algebraic tricks.

Although the argument may seem long, lasting as it does until page 66, it is really just a combination of several relatively short steps.

Step 1. For functions like $f(x) = x^2$, we were able to use algebra to expand expressions like $(x + h)^2$. In the present case we have the important "addition formula" from trigonometry:

$$\sin(x + y) = \sin x \cos y + \cos x \sin y.$$

Using this (with $y = h$) we can write

$$\sin'(x) = \lim_{h \to 0} \frac{\sin(x + h) - \sin x}{h}$$

$$= \lim_{h \to 0} \frac{\sin x \cos h + \cos x \sin h - \sin x}{h}$$

$$= \lim_{h \to 0} \left(\frac{\cos x \sin h}{h} + \frac{\sin x \cos h}{h} - \frac{\sin x}{h} \right).$$

Using simple facts about limits, briefly mentioned in Chapter 6, we can then write

$$(C) \quad \sin'(x) = \lim_{h \to 0} \left(\frac{\cos x \sin h}{h} \right) + \lim_{h \to 0} \left(\frac{\sin x \cos h}{h} - \frac{\sin x}{h} \right)$$

$$= \lim_{h \to 0} \left(\frac{\cos x \sin h}{h} \right) + \lim_{h \to 0} \left(\frac{\sin x (\cos h - 1)}{h} \right)$$

$$= \left[\lim_{h \to 0} \frac{\sin h}{h} \right] \cos x + \left[\lim_{h \to 0} \frac{\cos h - 1}{h} \right] \sin x.$$

Now this might very well not look like progress. Indeed, instead of *one* limit involving a fraction whose numerator and denominator both have the value 0 when $h = 0$, we now have *two* such limits! (Remember that $\cos 0 = 1$.)

But there's already one encouraging feature of this final expression. Naturally $\sin'(x)$ depends on x, as does the final expression of (C). Note, however, that the *limits* in this final expression *don't* depend on x. Once we have figured out the two specific limits

(L$_1$)
$$\lim_{h \to 0} \frac{\sin h}{h} = ?$$

(L$_2$)
$$\lim_{h \to 0} \frac{\cos h - 1}{h} = ?$$

(assuming that we *can* figure them out), we will know $\sin'(x)$ for all x.

Step 2a. Let's tackle (L$_1$) first. At this point we don't have any formulas or algebraic manipulations to save us; we're simply going to have to remember what everything means! The figure below shows, once again,

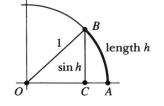

an angle of h (the circular arc $\overset{\frown}{AB}$ has length h), and $\sin h$ (the length of the perpendicular BC). Of course, we are really only interested in the situation when h is very small, so our picture should really be a very very tiny piece of the circle, with a corresponding very very short line BC.

But whether we draw it large or small, it certainly doesn't seem clear how we can say anything about the length of BC in terms of h, and for the moment it will be simpler to draw pictures with h fairly large, just so that we can see what we are talking about.

Remember that instead of determining the point B by the condition that the arc $\overset{\frown}{AB}$ has length h, we can just as well use the alternative condition (A) on page 61, that the sector OAB

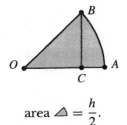

has area

$$\text{area } \triangle = \frac{h}{2}.$$

The figure below shows a shaded triangle $\triangle OAB$ within this sector. The

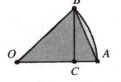

height of this triangle is the segment BC, which is just $\sin h$, while the base is the line OA, which has length 1, since this is a unit circle. So the area of the shaded triangle is

$$\text{area } \triangle = \frac{1}{2} \sin h \cdot 1 = \frac{\sin h}{2}.$$

Since we clearly have

$$\text{area } \triangle < \text{area } \triangle,$$

it follows that we have

$$\frac{\sin h}{2} < \frac{h}{2}.$$

Dividing by h we thus have

(a) $$\frac{\sin h}{h} < 1.$$

It wasn't really necessary to use areas to get this inequality. After all, $\sin h$ is the length of BC, and the length of this perpendicular is less than

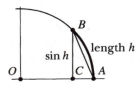

the length of the slanted line AB, while the length of this line AB is, in turn, less than the length h of the curved arc $\overset{\frown}{AB}$.

Step 2b. But areas will really be useful for our next piece of information. In the figure below, we will be looking at the triangle $\triangle OAB'$, which

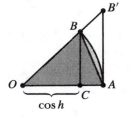

encloses our original sector. Since $\triangle OAB'$ is similar to $\triangle OCB$, we have

$$\frac{B'A}{OA} = \frac{BC}{OC} = \frac{\sin h}{\cos h},$$

and since OA has length 1, this means that

$$B'A = \frac{\sin h}{\cos h}.$$

Therefore the area of $\triangle OAB'$ is

$$\text{area } \triangle = \frac{1}{2} \cdot OA \cdot B'A = \frac{1}{2} \cdot 1 \cdot B'A = \frac{1}{2} \cdot \frac{\sin h}{\cos h}.$$

Since we clearly have

$$\text{area } \blacktriangle < \text{area } \triangle,$$

it follows that

$$\frac{h}{2} < \frac{1}{2} \cdot \frac{\sin h}{\cos h},$$

which gives

(b) $$\cos h < \frac{\sin h}{h}.$$

Step 3. Well, that took a bit of work, but now we are essentially done. Putting (a) and (b) together we have

(∗) $$\cos h < \frac{\sin h}{h} < 1.$$

Remember that when h is small, the number $\cos h$ is ***close to 1***. So (∗) shows that

$$\frac{\sin h}{h}$$

is between 1 and something which gets closer and closer to 1 as we make h smaller and smaller. This means that

$$\frac{\sin h}{h} \text{ itself}$$

must be getting closer and closer to 1 as we make h smaller and smaller. In other words we have established:

(L$_1$) $$\lim_{h \to 0} \frac{\sin h}{h} = 1.$$

Step 4. After all this work, you probably don't relish the idea of attacking the second limit (L$_2$). Fortunately, we no longer have to resort to geometric shenanigans. Having established (L$_1$), we can find the limit (L$_2$) with fairly standard algebraic trickery:

$$\lim_{h \to 0} \frac{\cos h - 1}{h} = \lim_{h \to 0} \frac{\cos h - 1}{h} \cdot \frac{\cos h + 1}{\cos h + 1}$$

$$= \lim_{h \to 0} \frac{(\cos h - 1)(\cos h + 1)}{h(\cos h + 1)}$$

$$= \lim_{h \to 0} \frac{\cos^2 h - 1}{h(\cos h + 1)}$$

$$= \lim_{h \to 0} \frac{-\sin^2 h}{h(\cos h + 1)} \quad \text{(remember page 21)}$$

$$= \lim_{h \to 0} \frac{\sin h}{h} \cdot \lim_{h \to 0} \frac{-\sin h}{\cos h + 1}.$$

Now the first limit is just the limit (L$_1$) that we have already found. How about the second limit? Well, as so often happens, the numerator has the value 0 when $h = 0$; but in this case the denominator *doesn't!* When h is close to 0, the numerator is simply close to 2, so the whole fraction is close to 0. This means that the product limit is also 0. In other words, we have established:

(L$_2$) $$\lim_{h \to 0} \frac{\cos h - 1}{h} = 0.$$

Step 5. Finally, we simply stick the limits (L$_1$) and (L$_2$) back into (C) on page 62 and we obtain the desired formula

$$\boxed{\sin'(x) = \cos x.}$$

Although the steps leading up to it might be considered rather complicated, you have to admit that the final result itself is about the simplest formula one could hope for! Still, it might be a good idea to check that the formula really gives results that appear reasonable.

First of all, we have

$$\sin'(0) = \cos 0 = 1.$$

This means that the tangent line to the graph of $f(x) = \sin x$ at the point $(0,0)$ has slope 1, which seems reasonable from the graph. At the

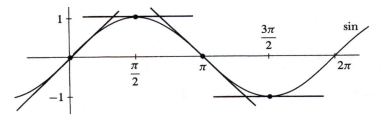

point $\pi/2$, a "peak" of the graph, we have

$$\sin'(\pi/2) = \cos \pi/2 = 0,$$

while at π we have

$$\sin'(\pi) = \cos \pi = -1,$$

and then at $3\pi/2$, a "valley," we once again have

$$\sin'(3\pi/2) = \cos(3\pi/2) = 0.$$

When the graphs of sin and cos are presented together, you should be able to convince yourself that cos has at least the right basic appearance

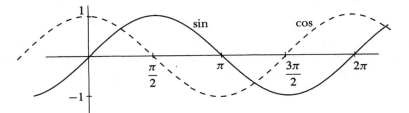

for the derivative sin': that is, cos is positive at the places where sin' should be positive, negative at the places where sin' should be negative, and zero at the appropriate places.

Having discovered the derivative of the sin function, it is natural to ask for the derivative of the cos function. A straightforward procedure is simply to start with the definition

$$\cos'(x) = \lim_{h \to 0} \frac{\cos(x + h) - \cos x}{h},$$

and use another "addition formula" from trigonometry:

$$\cos(x + y) = \cos x \cos y - \sin x \sin y.$$

We quickly find that

$$\cos'(x) = \left(\lim_{h \to 0} \frac{\sin h}{h} \right) \cdot (-\sin x) + \left(\lim_{h \to 0} \frac{\cos h - 1}{h} \right) \cdot \cos x,$$

and the limits (L_1) and (L_2) thus lead directly to the result

$$\boxed{\cos'(x) = -\sin x.}$$

Remember the minus sign here—it's important.

Once again, when the graphs of cos and sin are presented together, you should be able to convince yourself that the function $f(x) = -\sin x$

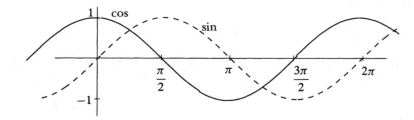

has at least the right basic appearance for the derivative cos': that is, sin is negative at the places where cos' should be positive, positive at the places where cos' should be negative, and zero at the appropriate places.

10   *The view from on high.*

S o far, our journey has probably seemed like a continuous uphill climb. It would be comforting to learn that we've reached the peak, and that from now on we can gleefully anticipate that it's all going to be downhill.

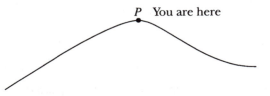

P   You are here

Or, even if that isn't in the cards, perhaps we can at least anticipate that it will be downhill for a while!

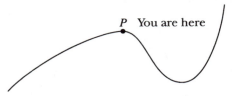

P   You are here

If we imagine this route of our journey to be the graph of the function f, the peak P is called a **maximum point** of f in the first case; in the second case, where P is only a maximum point nearby, we speak of a **local maximum point**.

Actually, we usually focus our attention simply on the first coordinate x_0 of the point P: we say that f **has a maximum at** x_0, or that x_0 is a

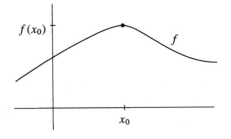

maximum point of f if

$$f(x_0) > f(x) \quad \text{for all } x \neq x_0.$$

69

In the second case, where we can at least choose numbers $a < x_0 < b$

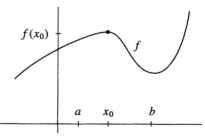

so that we have

$$f(x_0) > f(x) \quad \text{for all } x \text{ satisfying } a < x < b \text{ and } x \neq x_0,$$

we say that f **has a local maximum at** x_0, or that x_0 is a **local maximum point** of f. The value $f(x_0)$ itself is called a **maximum value** or a **local maximum value** of f.

Similarly, we can define a **minimum** or **local minimum** of f (and a **minimum value** or **local minimum value**).

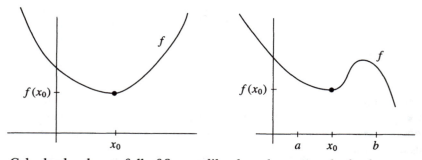

Calculus books are full of figures like these because calculus has something simple and striking to say about (local) maximum and minimum points: at such a point, the tangent line to the graph is *horizontal*; in other words, $f'(x_0) = 0$.

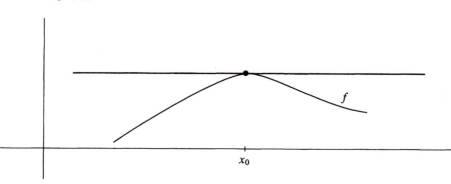

The first time I read this assertion in a calculus book I was somewhat perturbed, but not because I didn't believe it. In fact, I found the assertion very easy to believe—it certainly *looks* true in drawings. But mathematicians are so damned persnickety, always raising some subtle point that you've overlooked, and always demanding proof for apparently obvious assertions, that I didn't feel very secure basing my acceptance on a few pictures. And there are infinitely many different pictures one could draw

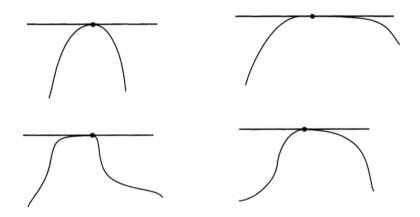

to illustrate the same point. How can it really be clear that the tangent line will be horizontal in all cases?

Actually, the idea that we must have a horizontal tangent line at a maximum or minimum point has already appeared, disguised, as a problem at the end of Chapter 4, where we considered the tangent line of the

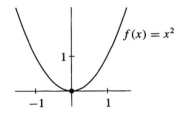

graph of $f(x) = x^2$ through $(0, 0)$, which happens to be its minimum point. At that time we hadn't done any calculations at all, so you had to imagine looking at the graph through a high-powered magnifying glass, and realize that the tangent line would be the x-axis.

This was probably pictorially clear (especially since the graph is symmetric about the y-axis, so that its magnified image must be symmetric also). More important, however, this conclusion follows easily from the fact that (0, 0) is a minimum point. After all, if the magnified graph instead looked like

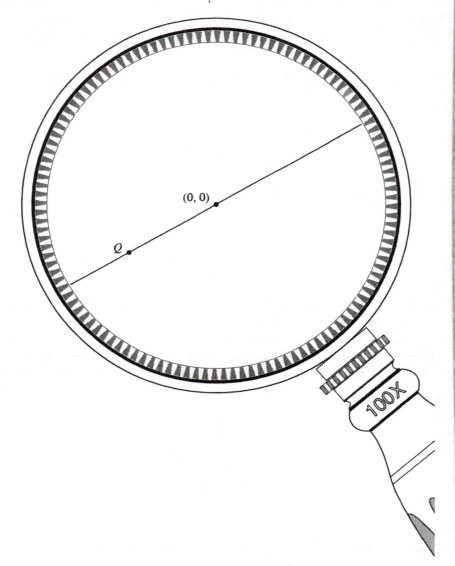

then the point (0, 0) wouldn't be a minimum point, because at points like Q the function would have a smaller value. And by the same token,

the magnified graph couldn't look like

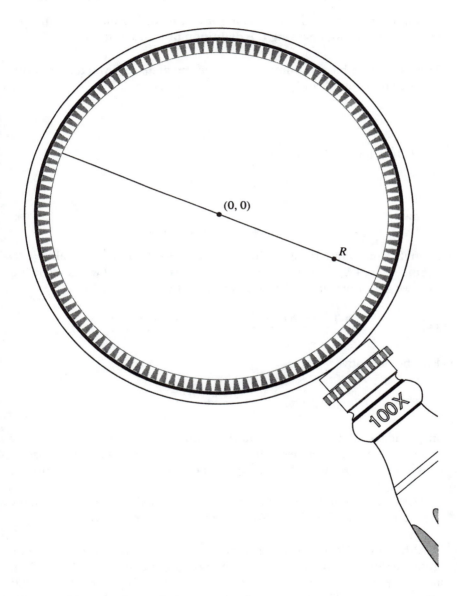

since then the function would have a smaller value at points like R. (Since the magnified graph isn't exactly a straight line, it would be more precise to say that there would be a point on the graph very close to Q or R where the function would have a smaller value.)

Consequently, the magnified graph must look like a horizontal line.

Moreover, it's easy to turn this argument involving magnified images into a simple proof involving the definition in terms of limits:

Theorem. If f has a minimum or maximum at x_0 (or even a local minimum or maximum) and f is differentiable at x_0, then

$$f'(x_0) = 0.$$

Proof: If f has a minimum at x_0, then

$$f(x_0 + h) > f(x_0)$$

so that

$$f(x_0 + h) - f(x) > 0.$$

Even if f has only a local minimum at x_0, this still holds for small $h \neq 0$.

Since $f(x_0 + h) - f(x)$ is always positive, this means that when we divide by h we get different results, depending on the sign of h:

$$(*) \qquad \frac{f(x_0 + h) - f(x)}{h} \quad \text{is} \quad \begin{array}{ll} > 0 & \text{if } h > 0 \\ < 0 & \text{if } h < 0. \end{array}$$

But the derivative $f'(x_0)$ is

$$f'(x_0) = \lim_{h \to 0} \frac{f(x_0 + h) - f(x)}{h};$$

in particular, the fraction on the right must be close to $f'(x_0)$ both for $h > 0$ and $h < 0$. And according to $(*)$, this means that $f'(x_0)$ must be close to both positive and negative numbers. The only way this is possible is if $f'(x_0) = 0$.

The proof when f has a maximum or local maximum at x_0 is virtually the same, except that all the inequalities are reversed. **Q.E.D.**

Although we've usually been careful to sidestep details, there are some good reasons for including this formal theorem and proof. First of all, it's always nice to have a proof that's easy, especially when we've held out the promise of a downhill portion on our journey. And it's certainly gratifying that something intuitively obvious turns out to have an easy proof—as we'll see later, we won't always be so lucky.

But there's another reason for emphasizing this concise and elegant observation: it is the basis for a whole slew of inelegant and tedious problems that you will undoubtedly have to reckon with in your calculus course.

As an example of such a problem, suppose that we want to construct a tin can of some fixed volume V. For example, we might want V to be exactly 14 fluid oz. or $13\frac{1}{4}$ oz. or $14\frac{3}{4}$ oz. (just to choose some examples that I've found on the grocery shelf). If the ends of the can have radius r, and thus area πr^2, and the can has height h, then the volume is given by the formula

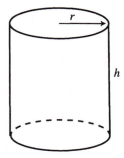

$$\text{volume} = \pi r^2 h.$$

Since we want to have

$$\pi r^2 h = V,$$

this means that once we choose the radius r, the height h is determined by the formula

$$(*) \qquad\qquad h = \frac{V}{\pi r^2}.$$

Now suppose that we want the surface area of this cylinder to be as small as possible, so that we will use as little tin as possible. The surface area consists of three pieces: the top and bottom circular pieces, each of which

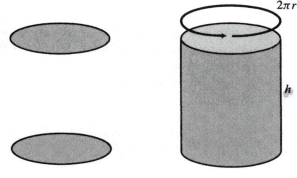

has area πr^2, and the cylindrical side. The height of this cylindrical side is just h, while its circumference is just the circumference, $2\pi r$, of the circular pieces. So the total area $A(r)$ is

$$2\pi r^2 + 2\pi r \cdot h,$$

or, according to (∗),

$$A(r) = 2\pi r^2 + 2\pi r \cdot \frac{V}{\pi r^2}$$
$$= 2\pi r^2 + \frac{2V}{r}.$$

So the problem is to choose r so that $A(r)$ is smallest.

Now, deep down, you probably don't care all that much about producing cans with the smallest amount of tin (politically incorrect though this thought might be). As a matter of fact, the makers of tin cans don't care either, because the price of the tin (actually steel with a thin tin coating) is negligible compared to all the other costs involved (and they probably don't worry about being politically correct, either). It's a lot simpler to have machines that produce cans with just a few fixed widths, but of variable height, especially since it's important to make cans that fit easily in packing crates and on grocery shelves.

But this is at least an example of a kind of problem that you, or some one, might actually want to solve some day; and since such problems can be solved by calculus, you're pretty much doomed to encounter them. So, although this is only a "toy" problem, we want to see how calculus helps us solve it.

According to the Theorem, at the point where the function

$$A(r) = 2\pi r^2 + \frac{2V}{r}$$

has a minimum we will have to have $A' = 0$. Finding A' should be no problem by this stage of your calculus course. In fact, in Chapter 5 we already found the derivatives of $f(r) = r^2$ and $f(r) = 1/r$, so we just need to use simple rules about sums of derivatives and products by constants. We find that

$$A'(r) = \frac{dA}{dr} = (2\pi) \cdot 2r + (2V) \cdot -\frac{1}{r^2}$$
$$= 4\pi r - \frac{2V}{r^2}.$$

So $A'(r) = 0$ is equivalent to

$$4\pi r = \frac{2V}{r^2},$$
$$2\pi r^3 = V,$$
$$r^3 = \frac{V}{2\pi}.$$

Thus, we obtain a can of volume V with the smallest surface area when we choose

$$r = \sqrt[3]{V/2\pi}.$$

This illustrates the basic idea behind the application of the Theorem, although in practice considerably more work may be involved: there may be more than one point where the derivative is 0, we will have to distinguish between maximum and minimum points, etc.

Moreover, as already hinted, there may be some complications lurking in the wings. In fact, we just have to look at the function $f(x) = x^3$, with $f'(0) = 0$, to realize that the converse of the theorem **isn't** true: if

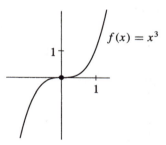

f has a (local) maximum or minimum at x, then $f'(x)$ must be 0, but just because $f'(x) = 0$, we **can't** conclude that x is a local maximum or minimum point.

Fortunately, this complication doesn't detract from the value of the Theorem in solving these "max-min" problems—it's still true that we can find all possibilities for a maximum or minimum by finding all x for which $f'(x) = 0$; the only problem is that all such x are just *candidates*, which then have to be scrutinized more carefully to determine their suitability.

As a matter of fact, instead of simply determining the relative maximum and minimum points of a function f, we can use the derivative to get a general idea of the shape of its graph.

To do this, we rely on two simple principles:

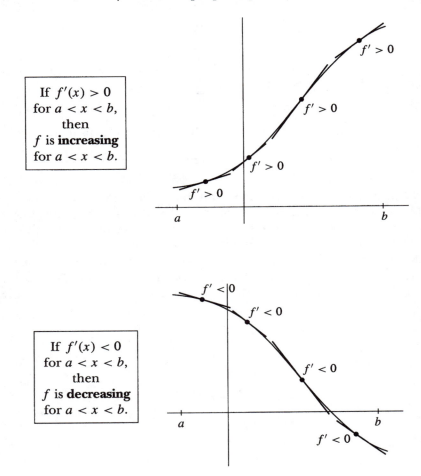

If $f'(x) > 0$
for $a < x < b$,
then
f is **increasing**
for $a < x < b$.

If $f'(x) < 0$
for $a < x < b$,
then
f is **decreasing**
for $a < x < b$.

By contrast, we also have the simple principle:

If $f'(x) = 0$
for $a < x < b$,
then
f is **constant**
for $a < x < b$.

For example, let's consider once again the function

$$A(x) = 2\pi x^2 + \frac{2V}{x},$$

with

$$A'(x) = \frac{dA}{dx} = 4\pi x - \frac{2V}{x^2}$$
$$= \frac{2(2\pi x^3 - V)}{x^2}.$$

We've already seen that $A'(x) = 0$ only for $x = \sqrt[3]{V/2\pi}$. The final formula for $A'(x)$ shows that $A'(x) > 0$ precisely when

$$2\pi x^3 - V > 0, \quad \text{i.e.,} \quad x^3 > V/2\pi.$$

If we temporarily consider only $x > 0$ (the only values that were of interest for our original problem), this is equivalent to $x > \sqrt[3]{V/2\pi}$. So A is increasing for $x > \sqrt[3]{V/2\pi}$, but decreasing from 0 to $\sqrt[3]{V/2\pi}$. In the figure below we have sketched the graph. The graph of $f(x) = 2\pi x^2$ was

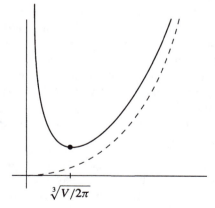

$$\sqrt[3]{V/2\pi}$$

sketched as a dashed guide, since for large x the value of $A(x)$ is just a bit more than $2\pi x^2$; for small x, on the other hand, it is the $2V/x$ part that is most important.

For $x < 0$ we have $x^3 < 0$, so certainly $A'(x) < 0$. Thus, the function A must be decreasing for $x < 0$; a little thought shows that the entire graph must look something like the following.

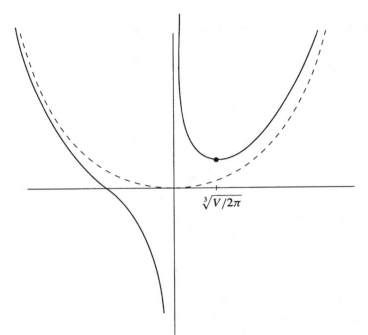

$$\sqrt[3]{V/2\pi}$$

If this picture looks familiar, it's because we graphed the function $f(x) = x^2 + 1/x$ in Chapter 3. That graph was made by a computer program that plotted many points, so all the values are very precise. As you can see, however, knowledge of the derivative allows us to sketch the general shape of the graph with hardly any calculation of specific points.

Our example also shows that even a relatively simple function can have some interesting features in its graph. That's why "graph-sketching" turns out to be a fairly prominent feature of all calculus courses, to which a considerable amount of time is devoted. With our typical insouciance, we will leave all the gory details to your course, and rest content with an understanding of the main ideas.

It still remains to justify the principles that we stated on page 78, but perhaps that doesn't seem like a particularly urgent task. Surely these principles are obvious, aren't they? And if they're not, is this something we really want to worry about?

11   *Scaling vertical walls. We watch the experts at play.*

My itinerary through the Belgian cities of Bruges, Ghent, Brussels and Namur finally landed me in the mountainous Ardennes region. Camping here for a couple of days, I managed to do some rock climbing, an interesting diversion, and of course a satisfyingly macho thing to do.

But I was quite content to join the other tourists watching the real climbers scaling the tall, nearly vertical cliffs at the edge of the river, cautiously seeking secure holds to move a bit further up.

Like rock climbing, which seems to have quite limited appeal (I've given it up long ago), proving theorems is not one of the most popular of sports. Even people who exult in the power of calculation often prefer to leave mathematical proofs to the sort of people who, ... well, to the sort who enjoyed geometry in high school. Although the proof in the last chapter was easy, many turn out to be much trickier. So in this chapter you may wish to adopt the tourist's observational role.

Even if you do like proving theorems, attempting to prove the principles stated on page 78 will probably seem like climbing straight up a smooth vertical wall—there just doesn't seem to be any way to get a good hold on the problem. Consider, for example, the third, and simplest principle:

If $f'(x) = 0$ for $a < x < b$, then f is constant for $a < x < b$.

It's hard to imagine how this could possibly be false. In fact, any non-constant function f that you *do* imagine will certainly have places where the derivative isn't 0. The problem is, we can only imagine a finite number of functions, not all of them, and perhaps our imagination simply isn't good enough; failure to imagine how the statement could be false does not, alas, constitute a mathematical proof that it is true (the reason, perhaps, why mathematicians often aren't welcomed on juries).

Before worrying about proofs, we first want to point out one simple consequence of this principle: If f and g have the same derivative everywhere, that is, $f'(x) = g'(x)$ for all x, then $g(x) = f(x) + C$ for some constant C, so that the graph of g is parallel to that of f. To see

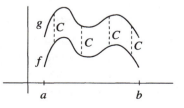

this, we simply consider the function $h(x) = g(x) - f(x)$. Then we have $h'(x) = g'(x) - f'(x) = 0$. So, according to the principle, h is a constant, $h(x) = C$ for all x, which means that $g(x) - f(x) = C$, or $g(x) = f(x) + C$.

If we are bent on providing proofs for our basic principles, then it turns out, just as with rock climbing, that our goal must be approached from a rather circuitous path. To begin with, let's consider a function f which satisfies the condition

$$f(a) = f(b)$$

for two numbers $a < b$. We'll want to assume that f is continuous at x for all $a \le x \le b$ (continuous functions are just about the only kind we've ever been considering), and we also want to assume that f is differentiable at x for $a < x < b$ (we don't particularly care whether it's differentiable at a or b).

First let's consider a point x which is a maximum for f, on the interval from a to b. Our simple Theorem from the previous chapter says that we

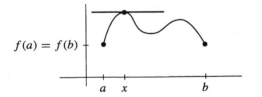

have $f'(x) = 0$, so we've found a place where the derivative of f is 0.

Actually, this picture might be wrong—because the graph of f might never go above the value $f(a) = f(b)$. But we can also consider a point x

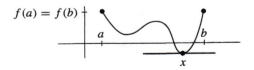

which is a *minimum* for f, and then the same conclusion holds: $f'(x) = 0$.

Finally, can both of these pictures be wrong? That is, can it happen that the graph of f never goes above the value $f(a) = f(b)$ and also never goes below it? Sure it can, but only if f has the constant value $f(a) = f(b)$, and in that case we have $f'(x) = 0$ for *any* x with $a < x < b$.

Oddly enough, these simple observations have been accorded the status of a Theorem (and odder still, it is named after a mathematician who had nothing to do with calculus):

Rolle's Theorem. Suppose that f is continuous at x for $a \leq x \leq b$ and differentiable at x for $a < x < b$ and suppose moreover that

$$f(a) = f(b).$$

Then there is some x satisfying $a < x < b$ for which we have

$$f'(x) = 0.$$

The figure below shows the graph of a function f, defined for $a \leq x \leq b$, where we certainly don't have $f(a) = f(b)$, so that Rolle's Theorem

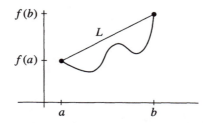

doesn't apply. The straight line L that passes through the two end-points $(a, f(a))$ and $(b, f(b))$ of our curve has slope

$$\text{slope of } L = \frac{f(b) - f(a)}{b - a}.$$

Instead of looking for a maximum or minimum of f, we will instead look for the point P where the graph of f is *furthest from L*. In the picture

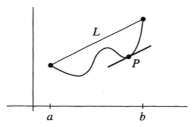

it looks as if the tangent line at this point is *parallel* to L.

Indeed, if we place our magnifying glass over the graph of f at the point P, we can't get a picture like

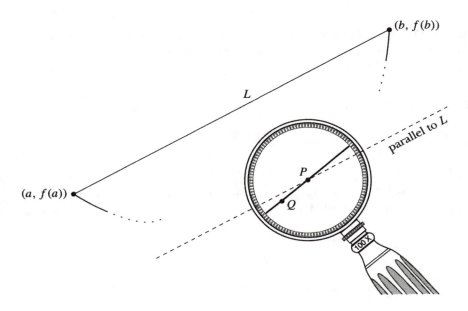

because then a point like Q would be further from L than P is.

So it really must be true that the tangent line at $P = (x, f(x))$ is parallel to L, which means that

$$f'(x) = \text{slope of } L = \frac{f(b) - f(a)}{b - a}.$$

If we think of $f(t)$ as the position of some object at time t, then this slope represents the average velocity of this object from the time a to the time b. In other words, we have found an x with $a < x < b$ so that the **(instantaneous) velocity** $f'(x)$ is exactly equal to the average velocity, or "mean" velocity, from time a to time b. This result is know as:

The Mean Value Theorem. If f is continuous at x for $a \le x \le b$ and differentiable at x for $a < x < b$, then there is some x with $a < x < b$ for which we have

$$f'(x) = \frac{f(b) - f(a)}{b - a}.$$

Our magnifying glass argument for the Mean Value Theorem might seem a tad unsatisfactory, but that's something we'll worry about later. The important thing is simply to remember the statement, and understand the meaning, of this result, which turns out to be one of the most important theoretical results in calculus. If some one ever asks you a theoretical-sounding question about calculus and you don't know the answer, try saying "It follows from the Mean Value Theorem" and you'll be right 90% of the time (the other 10% of the time you may end up looking pretty silly).

Just to prove this, we'll now establish the principles which seemed so intractable before.

Theorem. If $f'(x) > 0$ for $a < x < b$, then f is increasing from a to b. If $f'(x) < 0$ for $a < x < b$, then f is decreasing from a to b. And if $f'(x) = 0$ for $a < x < b$, then f is constant for $a < x < b$.

Proof: Consider any two points x_0 and x_1 between a and b with

$$x_0 < x_1.$$

We want to compare the values of $f(x_0)$ and $f(x_1)$, and to do this, we are going to apply the Mean Value Theorem *to the interval from x_0 to x_1*.

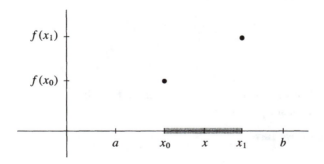

It tells us that

$$\frac{f(x_1) - f(x_0)}{x_1 - x_0} = f'(x),$$

or

$$(*)\qquad f(x_1) - f(x_0) = f'(x) \cdot (x_1 - x_0)$$

for some x between x_0 and x_1, and thus also between a and b.

Now if $f'(x)$ is always positive, then $(*)$ shows that $f(x_1) - f(x_0)$ is also positive (the factor $x_1 - x_0$ is positive, since we selected $x_0 < x_1$). In other words, if $f'(x)$ is always positive, then $f(x_1) > f(x_0)$ for $x_1 > x_0$, so that f is increasing.

Similarly, if $f'(x)$ is always negative, then $f(x_1) - f(x_0)$ is also negative, so that $f(x_1) < f(x_0)$. Thus, f is decreasing.

Finally, if $f'(x)$ is always 0, then $f(x_1) - f(x_0) = 0$, i.e., $f(x_0) = f(x_1)$ for any two points x_0, x_1 between a and b, so that the function f is constant for $a < x < b$. **Q.E.D.**

After this success, you might crave a somewhat more analytic proof of the Mean Value Theorem, not relying on pictures of magnifying glasses. Here, for the more adventurous tourist, is the standard argument, which derives the Mean Value Theorem directly from Rolle's Theorem. Instead of finding the desired point by a geometric criterion, we just use some slightly tricky algebraic manipulations.

Proof of the Mean Value Theorem: We will use m for the ratio

$$m = \frac{f(b) - f(a)}{b - a}.$$

Consider the function g defined in terms of f by

$$g(x) = f(x) - m(x - a)$$
$$= f(x) - \left[\frac{f(b) - f(a)}{b - a} \right] (x - a).$$

Although it may look a little complicated, we have simply cooked up g so that it has a simple property.

First of all, we have
$$g(a) = f(a).$$

Second of all,
$$g(b) = f(b) - m(b - a)$$
$$= f(b) - \left[\frac{f(b) - f(a)}{b - a} \right] (b - a)$$
$$= f(b) - [f(b) - f(a)]$$
$$= f(a).$$

In other words, we have

(1) $$g(b) = g(a).$$

Also note in the definition

$$g(x) = f(x) - m(x - a) = f(x) - mx + ma,$$

that m is just a constant, so that we have

(2) $$g'(x) = f'(x) - m.$$

Now because of (1), Rolle's Theorem applies to g, so we have

$$g'(x) = 0$$

for some x. But according to (2) this means that

$$f'(x) - m = 0,$$

which is just what we wanted to prove. **Q.E.D.**

12   *Ausfart. The Integral.*

T he most embarrassing event of my summer as an innocent abroad occurred shortly after I entered Germany. I had obtained a road map of the country at the border, and a lift by the friendly driver of an 18-wheeler. Since he spoke only German, and I spoke none, I decided that I would simply disembark at some interesting sounding city, and make my way from there.

Accordingly, when an appropriate sign appeared on the road, I motioned my intentions, alighted from the truck, waved good-bye, took out my map, and started searching for the city of *Ausfart.* This quest was doomed from the start, since—as I later found out—*Ausfart* is simply the German word for *exit.*

However, the tiny German city where I had landed turned out to be as good a place to begin as any other. When you're hitchhiking in a foreign land it doesn't really matter if you know where you're going—any place will probably end up being interesting.

That is the spirit that is going to guide us in this chapter. The road we've been following has led us to a summit. Now we're going to veer off in an entirely different direction, and see if the new path also takes us somewhere interesting.

The graph of the function $f(x) = x^2$ has been a natural starting point on all our little outings. It is, on the one hand, just about the simplest function after the linear functions, so it was natural to consider it first when we investigated derivatives. Since it is also a function that arises in the most elementary physics, this led us naturally to a discussion of velocity.

The graph of this function, a parabola, is also one of the simplest geometric shapes, after polygons and circles, so it was gratifying that we could use the derivative to find tangent lines to the graph.

But so far we've neglected one of the most important concepts connected with geometric figures—their area. It is only natural to try to find a formula for the area of a region like the following, bounded by the

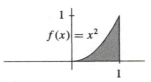

graph of the function $f(x) = x^2$, the x-axis, and the vertical line $x = 1$.

At this point we ought to take stock and remember what sort of tools we have for determining areas. The basic rules state that the area of a rectangle is the base times the height, while the area of a triangle is half that product:

$$\text{area} = \frac{1}{2}h \cdot b$$

And, as every school child knows, the area of a circle of radius r is πr^2. This is one of those rare cases where the traditional appeal to knowledgeable school children might be justified, for the formula is usually drummed into your head long before you get to a geometry course. Indeed, it might be a good idea to recall how we could actually go about establishing such a formula.

The figure below shows a circle surrounded by a polygon P made up of 10 congruent triangles, one of which has been shaded. The height of

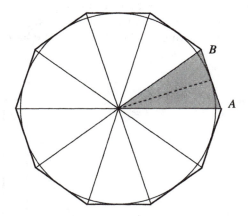

the shaded triangle, that is, the length of the dashed line, is simply the radius r of the circle, while its base is the length AB. So the triangle has area

$$\frac{1}{2}r \cdot AB.$$

Adding this over all triangles, we see that

$$\text{area of } P = \frac{1}{2}r \cdot \text{perimeter of } P.$$

The same reasoning applies if we consider an analogous 100-sided polygon, or a 1,000-sided polygon, ... For all such polygons P we have

$$(*) \qquad \text{area of } P = \frac{1}{2}r \cdot \text{perimeter of } P.$$

As we choose polygons with a larger and larger number of sides, the areas on the left side of equation $(*)$ will come closer and closer to the area of the circle. On the other hand, the perimeters of P on the right side of the equation will come closer and closer to the circumference of the circle. Consequently, we must have

$$\text{area of the circle} = \frac{1}{2}r \cdot \text{circumference of the circle}$$

$$= \frac{1}{2}r \cdot 2\pi r = \pi r^2.$$

Thus the mysterious number π in the formula for the circumference of the circle also shows up in the formula for its area.

The sort of argument used here, approximating the area of the circle by the areas of simpler figures, suggests a similar procedure for other areas. For example, in the figure below, we have surrounded a region bounded by a parabola with 10 rectangles. The length of the base of each rectangle

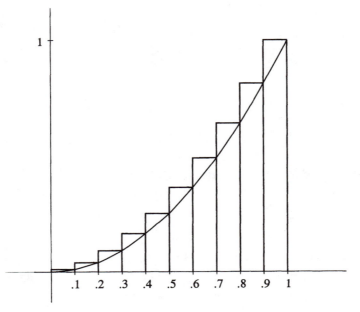

is simply .1, while the heights of the rectangles are given by

$$\text{height of rectangle 1} = .1^2$$
$$\text{height of rectangle 2} = .2^2$$
$$\text{height of rectangle 3} = .3^2$$
$$\cdots$$
$$\text{height of rectangle 10} = 1^2.$$

Consequently, the total area of the rectangles is

(S_{10}) $.1 \times (.1^2 + .2^2 + .3^2 + .4^2 + .5^2 + .6^2 + .7^2 + .8^2 + .9^2 + 1^2).$

Figuring this out is obviously a lot easier with a calculator, or even better, a programmable calculator, or better yet, an appropriate computer program. But no matter how we compute it, the answer won't be all that close to the area of the region that we are interested in because the rectangles extend considerably beyond the region.

We might hope to get a better answer by trying 100 rectangles, which will hug much closer to the region under the parabola. This is pretty hard to draw, but not so hard to write down. We now have 100 rectangles, each with a base of length .01, and we want to consider the sum

(S_{100}) $.01 \times (.01^2 + .02^2 + .03^2 + \cdots + .98^2 + .99^2 + 1^2).$

Now, of course, we'll really need a programmable calculator or computer, and we'll certainly need them for 1,000 rectangles. I got the following answers:

number of rectangles	total area
10	.385
100	.33835
1,000	.3338335.

But when I tried 10,000 rectangles the answer was given in the form 333.409×10^{-3} $(= .333409\ldots)$, indicating that the round-off errors from all the calculations prevented the answer from being given with complete accuracy.

You might or might not perceive a pattern here, but in any case it's clear that this method doesn't seem very efficient!

Instead of choosing a specific number of rectangles, like 10, or 100, or 1,000, ... , let's simply choose n rectangles, as in the figure below. Now

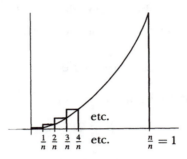

the length of the base of each rectangle is $\frac{1}{n}$, while the heights of the rectangles are given by

$$\text{height of rectangle } 1 = \left(\tfrac{1}{n}\right)^2$$

$$\text{height of rectangle } 2 = \left(\tfrac{2}{n}\right)^2$$

$$\text{height of rectangle } 3 = \left(\tfrac{3}{n}\right)^2$$

$$\cdots$$

$$\text{height of rectangle } n = \left(\tfrac{n}{n}\right)^2.$$

Consequently, the total area of the rectangles is

$$\frac{1}{n} \cdot \left[\left(\frac{1}{n}\right)^2 + \left(\frac{2}{n}\right)^2 + \left(\frac{3}{n}\right)^3 + \cdots + \left(\frac{n}{n}\right)^2 \right],$$

which can also be written in the form

$$\frac{1}{n} \cdot \left[\frac{1^2}{n^2} + \frac{2^2}{n^2} + \frac{3^2}{n^2} + \cdots + \frac{n^2}{n^2} \right],$$

or simply

$$(S_n) \qquad \frac{1}{n^3} \cdot \left[1^2 + 2^2 + 3^2 + \cdots + n^2 \right].$$

Most of the complication of this formula has been hidden in the $\cdots$ of the sum $1^2 + 2^2 + 3^2 + \cdots + n^2$. But there also happens to be a very compact formula for this sum, namely

$$1^2 + 2^2 + 3^2 + \cdots + n^2 = \frac{n(n+1)(2n+1)}{6}.$$

You may wonder where in the world this formula came from! But let's not worry about that right now (it's not a formula that you're usually expected to know). We simply want to note that once we have this formula we can write the total area (S_n) of our rectangles as

$$\frac{1}{n^3} \cdot \frac{n(n+1)(2n+1)}{6} = \frac{1}{n^2} \cdot \frac{(n+1)(2n+1)}{6}$$

$$= \frac{1}{6} \cdot \frac{n+1}{n} \cdot \frac{2n+1}{n}.$$

The reason for writing our answer in this final form is that we can now see what happens when n is large: the first fraction $\frac{n+1}{n}$ is close to 1, while the second fraction $\frac{2n+1}{n}$ is close to 2. Consequently, the whole answer is close to

$$\frac{1}{6} \cdot 1 \cdot 2 = \frac{1}{3}.$$

And this means, finally, that the area of the figure on page 90 must actually be 1/3.

For the region bounded by the parabola and the vertical line $x = a$, the area turns out to be

$$\text{area} = \frac{a^3}{3}.$$

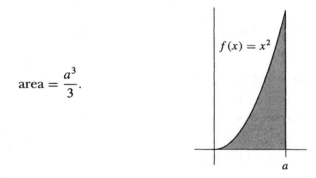

We could establish this in practically the same way, choosing n rectangles by means of the points $\frac{a}{n}, \frac{2a}{n}, \frac{3a}{n}, \ldots$. But we will forgo the exquisite agony of this calculation, since we've already made our point: It takes a lot of work just to find the area of one region! (And we would have been stymied without knowing a formula for $1^2 + 2^2 + 3^2 + \cdots + n^2$.)

One shudders to think what would be required to find the area of a region under the graph of $f(x) = x^3$ or $f(x) = x^4$, or even more complicated things like $f(x) = 1/(1 + x^2)$, let alone something like the graph of $f(x) = \sin x$!

At this point we must remember that when the mathematics gets tough, tough mathematicians introduce new notation.

We want to consider the area bounded by the graph of the function f, the x-axis, and the vertical lines $x = a$ and $x = b$. For convenience, we'll always assume that f is continuous.

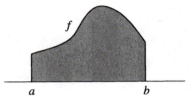

In our previous calculations we used rectangles whose bases all had the same length, but there's no particular reason for making this restriction. Let's divide the interval from a to b into n arbitrary size pieces by means of the points

$$a = t_0, t_1, \ldots, t_n = b.$$

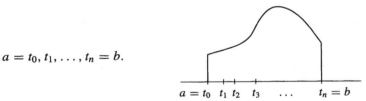

We have numbered the points in this particular way so that the length of the base of the

first rectangle will be $t_1 - t_0$
second rectangle will be $t_2 - t_1$
third rectangle will be $t_3 - t_2$

. . .

i^{th} rectangle will be $t_i - t_{i-1}$

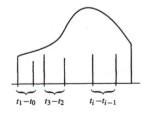

. . .

These numbers are often denoted by

$$\Delta x_1 = t_1 - t_0$$
$$\Delta x_2 = t_2 - t_1$$
$$\Delta x_3 = t_3 - t_2$$
$$\ldots$$
$$\Delta x_i = t_i - t_{i-1}$$

where the use of Δ is analogous to that on page 38.

There are several ways that we might choose the heights of the rectangles. For example, for the first (shaded) rectangle in the figure below we chose the maximum value of f over the first base, so that the rectangle protrudes above the graph.

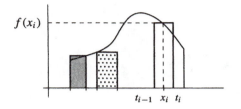

The second rectangle hasn't been drawn yet, but for the third (dotted) rectangle we chose the minimum value of f over the third base, so that this rectangle lies completely below the graph.

Finally, over the ith rectangle we've simply chosen an arbitrary point x_i between t_{i-1} and t_i, and made a rectangle with height $f(x_i)$, so that the rectangle lies partly below and partly above the graph.

Whether we choose the x_i to be arbitrary points between t_{i-1} and t_i, or the maximum or minimum points for f, we can now consider the sum

$$f(x_1)(t_1 - t_0) + f(x_2)(t_2 - t_1) + \cdots + f(x_n)(t_n - t_{n-1}),$$

which we can also write as

$$f(x_1)\Delta x_1 + f(x_2)\Delta x_2 + \cdots + f(x_n)\Delta x_n,$$

using the notation introduced on page 94.

Even more compactly, we can use the $\sum$ notation ("Sigma notation") to write this as

$$(*) \qquad \sum_{i=1}^{n} f(x_i)\Delta x_i$$

(read "the sum from $i = 1$ to n of $f(x_i)\Delta x_i$"). Here the $\sum$ indicates the sum of the quantities obtained from the next expression, $f(x_i)\Delta x_i$, by choosing $i = 1, 2, \ldots, n$.

When all the Δx_i are small, this represents the total area of rectangles all lying close to the graph of f. So we obtain the desired area by determining

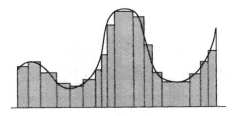

what these sums come closer and closer to as we make all the Δx_i smaller and smaller (a lot easier said than done!).

Note, however, that at places where f has negative values, the term

$f(x_i)\Delta x_i$ is *negative*.

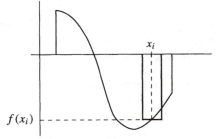

So regions below the x-axis contribute negative amounts. Thus we are actually considering a "signed" area—if f lies partly above the x-axis and partly below it, we are trying to compute the difference between the sum

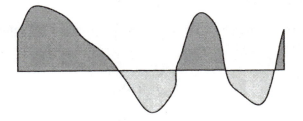

of the areas lying above the x-axis and the sum of the areas lying below it.

This signed area is called the *integral* of the function f from a to b, and is denoted by

$$\int_a^b f(x)\,dx.$$

This symbol is supposed to remind us of the sums

(*) $$\sum_{i=1}^n f(x_i)\Delta x_i$$

on page 95, and is derived from this expression in the same way that the

$$\frac{df(x)}{dx}$$ notation for derivatives is derived from the expression $\dfrac{\Delta f}{\Delta x}$

(see the top of page 39)—we replace all of the Δx_i by dx to indicate that we are considering what happens when all the Δx_i are very small. At the same time, the $\sum$ sign metamorphoses into an $\int$ sign (basically an elongated S) to indicate that we are summing over enormous numbers of very narrow rectangles.

Although the suggestive symbolism may be pleasing, the really important point about this notation is simply that

$$\int_a^b f(x)\,dx$$

contains all the essential information, namely the function f and the numbers a and b. It also contains the letter x, but this letter is merely a convenience for naming functions. For example,

$$\int_a^b x^2\,dx$$

means the number that the sums (*) are close to when we choose the function

$$f(x) = x^2;$$

thus, as we have so laboriously calculated,

$$\int_0^1 x^2\,dx = \frac{1}{3}.$$

More generally,

$$\int_0^a x^2 \, dx = \frac{a^3}{3},$$

as we mentioned on page 93, although we didn't go through with the details of the computation.

While it is customary to use the letter x in this equation, just as we customarily speak of the "x-axis," this isn't necessary, and one could just as well use another letter. For example, we can write

$$\int_0^a t^2 \, dt = \frac{a^3}{3}.$$

Notice that the left side of this equation simply refers to the function $f(t) = t^2$, which is just another way of naming the function $f(x) = x^2$. As in the definition "$f(x) = x^2$," the notation

$$\int_a^b f(x) \, dx$$

just uses the letter x as a "place-holder," that can be replaced by any other letter (except a or b or f of course!).

We've seen how hard it may be to calculate integrals, but at least we always know the value when $b = a$: For any function f we have

$$\int_a^a f(t) \, dt = 0.$$

(One value may not seem like much, but we might as well hang on tenaciously to every little bit of information we can get!)

There is also a geometrically obvious relation that it will be useful to have explicitly stated:

The "additivity property" of the integral. If $a < b < c$, then

(A) $$\int_a^c f(x) \, dx = \int_a^b f(x) \, dx + \int_b^c f(x) \, dx.$$

This equation simply expresses the geometric fact that in the figure below the area of the whole region is the sum of the areas of the two

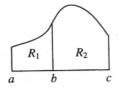

regions R_1 and R_2 (it really says something a bit more complicated than that, since the integral represents signed area).

As a typical use of this property, we can start with

$$\int_0^b x^2\,dx = \int_0^a x^2\,dx + \int_a^b x^2\,dx, \qquad 0 < a < b,$$

and then subtract $\int_0^a x^2\,dx$ from both sides to obtain

$$\int_a^b x^2\,dx = \int_0^b x^2\,dx - \int_0^a x^2\,dx.$$

Using the formula for $\int_0^b x^2\,dx$ (top of page 98), we then see that

$$\int_a^b x^2\,dx = \frac{b^3}{3} - \frac{a^3}{3}.$$

But we don't know much more than this. We've acquired some pretty notation—the symbols

$$\frac{df(x)}{dx} \quad \text{and} \quad \int_a^b f(x)\,dx$$

are the main mysterious notations from calculus—but that's about all. Our *Ausfart* has indeed lead us to an interesting new problem, but it isn't clear whether this problem will open up new vistas or will prove to be a dead end.

13

Twin Peaks. Meandering in this rarefied air,
we discover a bridge between two promontories.

U nlike our relatively smooth journey toward the derivative, our experience with the integral has been a rather arduous climb up steep
and rugged terrain. We were lead to introduce the notation

$$\int_a^b f(x)\,dx$$

for the "signed" area between the x-axis, the graph of the function f, and

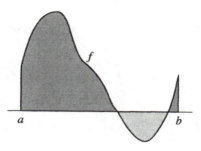

the vertical lines $x = a$ and $x = b$, with regions below the x-axis counting
as negative. But at the moment we have merely indicated how one might
compute this number in a special case, and a general computation seems
almost hopeless.

As we emphasized in the previous chapter, the letter x in this notation
is merely a "place-holder," and one can just as well use other letters, like

$$\int_a^b f(t)\,dt \qquad \text{or} \qquad \int_a^b f(u)\,du \qquad \text{etc.}$$

In this chapter we will almost always be using other letters, because instead
of simply looking at the number

$$\int_a^b f(t)\,dt,$$

we are going to be looking at the expression

$$\int_a^x f(t)\,dt.$$

Notice that here we have used t as the place-holder, while x has been used instead of b; we are doing this to emphasize that we are considering the area

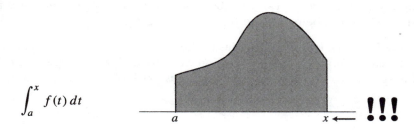

$$\int_a^x f(t)\,dt$$

for *all* numbers x, not just for a particular one. In other words, we are now all set up to consider the *function*

$$A(x) = \int_a^x f(t)\,dt.$$

Notice that we now have *two* place-holders. First of all, x is a place-holder for the definition of $A(x)$, giving us a sensible way of saying

$$A(\text{a number}) = \int_a^{\text{that number}} \quad \dots.$$

Moreover, within this $\int_a^{\text{that number}} \dots$ expression, the letter t is another place-holder, so that if we were considering an example like

$$A(x) = \int_a^x t^2\,dt,$$

then the $t^2\,dt$ would be indicating the function f with

$$f(\text{a number}) = \text{square of that number}.$$

Even apart from all this extra notational complexity, introducing the function A might seem like an exercise in futility. After all, knowing the function A amounts to knowing $A(x)$ for all x, and we generally cannot even figure out any particular value

$$A(b) = \int_a^b f(t)\, dt,$$

except for the special value $b = a$. Nevertheless—and this is a trick that mathematicians have been getting away with for hundreds of years—it might turn out that we can establish some important properties of this function without being able to compute it directly.

Now just what sort of properties of the function A might we expect to investigate? Well, . . . , this *is* a book on calculus, and we have spent a heck of long time on derivatives, so me might just try looking at the derivative

$$A'(x)$$

of our function A!

If the initial idea of introducing the function A seemed silly, the idea of considering the derivative $A'(x)$ of this function might seem utterly preposterous. On the one hand, we have a function A for which we might not be able to compute any values (and this function A isn't even a specific function, but depends on the function f that we selected to begin with). On the other hand, as was made painfully clear in Chapter 6, finding derivatives can require a great deal of work even for particular, relatively simple, functions. Trying to find the derivative of a function that we cannot even compute in the first place must be sheer madness!

But we're not going to let a little problem like that stop us, are we? Let's just go back to the definition (there's nothing else we can do)

$$A'(x) = \lim_{h \to 0} \frac{A(x + h) - A(x)}{h}.$$

Remember that A was defined by

$$A(x) = \int_a^x f(t)\, dt,$$

so that we also have

$$A(x + h) = \int_a^{x+h} f(t)\, dt.$$

Thus we are trying to compute

$$A'(x) = \lim_{h \to 0} \frac{A(x+h) - A(x)}{h}$$

$$= \lim_{h \to 0} \frac{\int_a^{x+h} f(t)\,dt - \int_a^x f(t)\,dt}{h}.$$

In this equation, x represents some particular number, which will remain fixed during the argument—it is h that we are going to be changing, letting it get closer and closer to 0. As is virtually always the case, we have a situation where the numerator and denominator both have the value 0 when $h = 0$, so some sort of sneaky manipulation will be required.

The first step, at least, isn't hard. The additivity property (**A**) of the integral (page 98) means that we can write

$$\int_a^{x+h} f(t)\,dt - \int_a^x f(t)\,dt = \int_x^{x+h} f(t)\,dt.$$

Consequently,

$$A'(x) = \lim_{h \to 0} \frac{\int_x^{x+h} f(t)\,dt}{h}.$$

But now that we've gotten a fairly simple looking symbolic expression, we are going to have to face the music and remember what this really means:

$$A'(x) = \lim_{h \to 0} \frac{\text{area} \underset{x \quad x+h}{}}{h}.$$

As always, the value of the limit depends only on what happens when h is very small, so our picture should really be

$$\lim_{h \to 0} \frac{\text{area} \underset{x \; x+h}{}}{h},$$

where we are taking the area of a very tiny sliver, and then dividing by the tiny number h.

Just as in Chapter 6, it hardly seems clear at first how we can possibly compare h to the area of this tiny sliver, especially since the area depends on the particular function f that we are working with.

The figure below shows a (somewhat expanded) sliver, together with a rectangle, having height m, lying below the graph, and another rectangle,

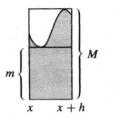

with height M, lying above the graph. The area of the rectangle below the graph is just $h \cdot m$, while the area of the rectangle above the graph is just $h \cdot M$. Since the area under the curve clearly lies between these two areas, we have

$$(*) \qquad\qquad h \cdot m < \text{area} \;\blacksquare\; < h \cdot M.$$

Remember that we are only interested in what happens when h is very small. In this case, because f is continuous at x, the minimum value m and the maximum value M of f on the small interval from x to $x + h$ must both be very close to the value $f(x)$ that f has at the endpoint x. So $(*)$ shows that for small h

area $\;\blacksquare\;$ must be close to $h \cdot f(x)$.

In picturesque terms, we might say that for small h our sliver is practically a rectangle with base h and height $f(x)$! But to be more precise, let us divide $(*)$ by h and write it in the form

$$m < \frac{\displaystyle\int_x^{x+h} f(t)\,dt}{h} < M.$$

Then we see that for small h

$$\frac{\displaystyle\int_x^{x+h} f(t)\,dt}{h} \qquad \text{must be close to } f(x).$$

This means that

$$A'(x) = \lim_{h \to 0} \frac{\displaystyle\int_x^{x+h} f(t)\,dt}{h} = f(x).$$

Thus, although we may not be able to compute $A(x)$, the derivative $A'(x)$ is easy to compute—it is just the value of the original function f at x!

In our pictures we have always been assuming that the graph of f lies above the x-axis. When f lies below the x-axis, basically the same arguments work, remembering that in this case all our areas must be taken as negative. When little details of this sort are attended to, our argument—written down just a wee bit more carefully—establishes

---

THE FUNDAMENTAL THEOREM OF CALCULUS. If f is continuous and the function A is defined by

$$A(x) = \int_a^x f(t)\,dt,$$

then for all x we have

$$A'(x) = f(x).$$

In this modern day and age, when everything and everyone is touted as the greatest, the best, and the most famous (at least for 15 minutes), you might think that the name attached to this theorem is just another bit of hoopla, not meant to be taken too seriously. Or is this theorem perhaps really so fundamental?

One argument that our theorem merits such a distinguished name is simply the fact that it establishes a connection between the two very different concepts of the integral and the derivative. But then again, that argument might not seem so persuasive when we recall that the integral was discovered only on a detour from our main route.

In fact, there's a much better reason for according the theorem such an exalted status. Until now the integral has been little more than a concept, to which we have associated some pretty notation, but which we haven't been able to compute in any reasonable way. But, as we'll soon see, with the appearance of the Fundamental Theorem of Calculus, it becomes a whole 'nother ball game.

Let's return to the area

$$\int_a^b t^2 \, dt$$

which began our discussion of integrals. The calculation of this integral was a major task. We not only had to derive the formula (S_n) on page 92, but we also relied on a special formula for the sum $1^2 + 2^2 + 3^2 + \cdots + n^2$. Now we're going to use the Fundamental Theorem to evaluate this integral with almost no computations at all!

Consistent with the strategy of this chapter, we simply introduce the function

$$A(x) = \int_a^x t^2 \, dt,$$

for which the Fundamental Theorem of Calculus tell us that

$$A'(x) = x^2.$$

Now the interesting thing about this little equation is that we can also name another, quite explicit, function with the same derivative:

$$\text{If } F(x) = \frac{x^3}{3}, \text{ then } F'(x) = x^2.$$

Thus A, the function which was defined in terms of areas, without an explicit formula, has exactly the same derivative as the function F with the very simple formula $F(x) = x^3/3$.

The fact that our mysterious function A and our simple explicit function F both have the same derivative doesn't necessarily mean that A and F are equal. But, as we pointed out on page 81, it does mean that A and F differ only by a constant: there is some number C such that

$$A(x) = F(x) + C \qquad \text{for all } x;$$

in other words,

(1) $$\int_a^x t^2 \, dt = \frac{x^3}{3} + C \qquad \text{for all } x.$$

We can easily figure out what C must be by choosing $x = a$ in (1): we get

$$0 = \int_a^a t^2 \, dt = \frac{a^3}{3} + C,$$

so that we must have

$$C = -\frac{a^3}{3}.$$

Finally, plugging this value of C back into (1) we obtain

$$\int_a^x t^2 \, dt = \frac{x^3}{3} - \frac{a^3}{3}.$$

Now that we've derived this formula, we might as well choose the specific value $x = b$ so that we can write

$$\int_a^b t^2\, dt = \frac{b^3}{3} - \frac{a^3}{3}.$$

Thus, we have figured out this area with hardly any computations at all; we simply had to know the derivative of $F(x) = x^3$.

By the way, notice that our formula gives us the area under the graph of $f(x) = x^2$ even for $a < 0 < b$: in this case, a^3 is negative, so the term $-a^3/3$ actually adds on another positive amount.

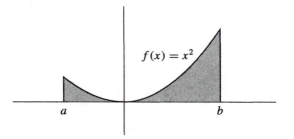

As you might imagine, there are now multitudes of other integrals for which we can give specific formulas. For example, we can easily compute

$$\int_a^b x^3\, dx,$$

using the fact that

the function $F(x) = \dfrac{x^4}{4}$ has derivative $F'(x) = x^3$.

For if we define

$$A(x) = \int_a^x t^3\, dt,$$

then

$$A'(x) = x^3 = F'(x),$$

so there is some number C such that

$$A(x) = F(x) + C,$$

that is to say,

$$\int_a^x t^3\, dt = \frac{x^4}{4} + C.$$

Choosing $x = a$ then gives $C = -a^4/4$, so that, finally,

$$\int_a^b t^3\, dt = \frac{b^4}{4} - \frac{a^4}{4}.$$

Notice that in this case we have to be a little more careful about the sign of a and/or b. If $a < 0 < b$, then the term $-a^4/4$ is negative, and represents the amount of area *below* the axis, which we are subtracting.

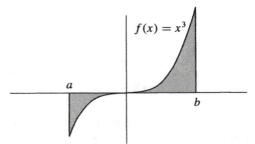

On the other hand, if $a < b < 0$, then $a^4 > b^4$, so

$$\frac{b^4}{4} - \frac{a^4}{4} < 0;$$

this number is the negative of the area beneath the graph of $f(x) = x^3$ between a and b.

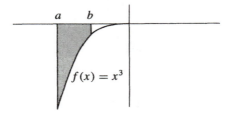

The method that worked in these two examples can be used in general: Suppose that we have a function F such that

$$F'(x) = f(x).$$ **(HYPOTHESIS)**

Then if we define

$$A(x) = \int_a^x f(t)\,dt,$$

the Fundamental Theorem tell us that

$$A'(x) = f(x).$$

So $A'(x) = F'(x)$ for all x, which means that for some constant number C we have

$$A(x) = F(x) + C,$$

i.e.,

$$\int_a^x f(t)\,dt = F(x) + C.$$

Choosing $x = a$ in this equation gives

$$C = -F(a),$$

so that we obtain finally

$$\int_a^b f(t)\,dt = F(b) - F(a).$$ **(CONCLUSION)**

The right side of this equation is sometimes written as

$$F\Big|_a^b \qquad \text{or} \qquad F(x)\Big|_{x=a}^{x=b},$$

so that we can write

$$\int_a^b f(t)\,dt = F(x)\Big|_{x=a}^{x=b} \qquad \text{for } F'(x) = f(x).$$

This equation is sometimes called the Second Fundamental Theorem of Calculus. Although just an immediate corollary of the Fundamental

Theorem, it is worth singling out in this way because of its crucial role in evaluating integrals.

It should be clear that we are now in a position to write down all sorts of fairly spectacular area formulas simply by "reversing" differentiation formulas. As a particularly interesting case, we can ask about

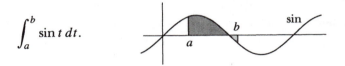

$$\int_a^b \sin t \, dt.$$

We can evaluate this integral because we can easily find a function whose derivative is sin: since $\cos'(x) = -\sin x$ (that $-$ sign is important!), we have

$$(-\cos)'(x) = \sin x.$$

Consequently,

$$\int_a^b \sin t \, dt = -\cos x \Big|_{x=a}^{x=b}$$
$$= -\cos b - (-\cos a)$$
$$= \cos a - \cos b.$$

To find the area under one arch of the sin curve, we simply have to

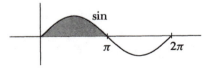

choose $a = 0$ and $b = \pi$, to obtain the surprising result

$$\int_0^\pi \sin t \, dt = \cos 0 - \cos \pi$$
$$= 1 - (-1)$$
$$= 2.$$

If we consider the signed area from 0 to 2π, then we would expect the

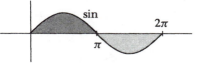

answer to be 0, since the second arch has the same area as the first, except that it lies below the x-axis, and thus counts as negative. And indeed we find that

$$\int_0^{2\pi} \sin t \, dt = \cos 0 - \cos 2\pi$$
$$= 1 - 1 = 0.$$

Finding the area under one arch of the cos function really shouldn't

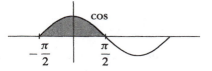

require any computations at all, since the graph of cos is just the graph of sin moved left by $\pi/2$. Consequently this area, which is given by the integral

$$\int_{-\pi/2}^{\pi/2} \cos t \, dt,$$

should be the same as the area under an arch of the sin curve. However, just to check, you ought to evaluate this integral using the Second Fundamental Theorem (the computations are even easier, since $\sin' = \cos$, and we don't have to worry about minus signs).

As these examples illustrate, although the Fundamental Theorems allow us to write down any number of formulas for integrals, this doesn't really solve the problem of evaluating integrals: when we are given a particular integral

$$\int_a^b f(t) \, dt$$

that we want to evaluate, we have to guess a function F for which we have $F'(x) = f(x)$.

For example, suppose we want to find

$$\int_a^b \sin t \cos t \, dt.$$

If we notice that

$$\frac{d}{dx}\sin^2 x = \frac{d}{dx}[(\sin x) \cdot (\sin x)]$$

$$= \sin x \cos x + \cos x \sin x$$
$$= 2 \sin x \cos x$$

(this requires the rule for the derivative of a product of functions), then we can conclude that

$$\int_a^b \sin t \cos t \, dt = \left. \frac{\sin^2 x}{2} \right|_{x=a}^{x=b}$$

$$= \frac{\sin^2 b}{2} - \frac{\sin^2 a}{2}.$$

But how would we be lead to notice this!? Unlike the case of differentiation, which requires only a methodical application of basic rules, finding a function F with $F' = f$ can be a real art.

Just how much time a calculus course devotes to this art depends, to some extent, on the purpose of the course. There are a few important rules, and some standard applications of these rules, that almost every course will cover. Once these standard rules are mastered, it may suffice to consult special "tables of integrals" that give many such formulas, or one may use special computer programs that evaluate integrals, or one may expand the repertoire of rules and tricks so that more and more integrals can be evaluated "by hand."

On your journey through calculus, the art of integration may turn out to be a short stay at a pleasant country inn, backbreaking labor in the vineyards, or a whirlwind tour through a high-tech museum. Alas, *The Hitchhiker's Guide* can't help you at this point! We'll simply have to wait until this episode is over, and then stick out our thumbs once again to continue on our way.

14 Approaching the end of our journey; the one with the most toys wins.

O ur attempts to find the area under the graphs of functions began with some very specific rectangles for the function $f(x) = x^2$, and then progressed to general sums for any function f,

$$(*) \qquad \sum_{i=1}^{n} f(x_i)\Delta x_i,$$

which in turn lead to the introduction of the notation $\int_a^b f(x)\,dx$. But our main success in evaluating integrals was by means of the Fundamental Theorem of Calculus, which we derived directly from simple properties of area, notably the additivity property (**A**) on page 98. So the introduction of the sums (*) might seem like an unnecessary embellishment.

Theoretical mathematicians hardly look at the matter this way, since for them the notion of area must be *defined*, and the most reasonable definition is by means of the sums (*).

But these sums also play another very important role. For it turns out that many quantities besides areas can be expressed in terms of such sums, and thus evaluated as integrals.

For example, suppose that we have a thin metal plate of width W, height H, and depth D, as shown in the figure on the left. Let's sup-

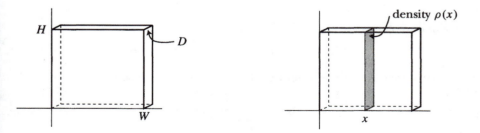

pose, moreover that the density of the metal in this plate varies, but only in a way that depends on x. In other words, for each x, there is a number $\rho(x)$ which is the density throughout the section at distance x from the origin, as illustrated in the figure on the right.

If the density ρ were a constant, then the weight of the metal plate would simply be the volume times ρ,

$$\text{weight} = D \cdot H \cdot W \cdot \rho.$$

The problem is to find a formula for the weight when the density varies with x.

In the figure below, we have picked points

$$0 = t_0, t_1, \ldots t_{i-1}, t_i, \ldots t_n = W,$$

together with points x_i between t_{i-1} and t_i. In the shaded portion, be-

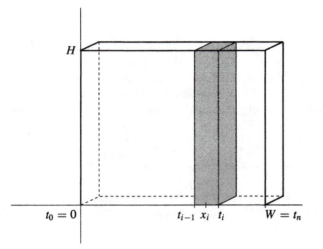

tween the sections at t_{i-1} and t_i, the density is very close to $\rho(x_i)$. So the weight of this portion of the metal plate is close to

$$D \cdot H \cdot (t_i - t_{i-1}) \cdot \rho(x_i),$$

which we can write as

$$D \cdot H \cdot \rho(x_i)\Delta x_i, \qquad \Delta x_i = t_i - t_{i-1}.$$

This means that the weight of the whole plate is close to

$$D \cdot H \cdot \sum_{i=1}^{n} \rho(x_i)\Delta x_i,$$

involving a sum of the form $(*)$ for the function ρ on the interval from 0 to W. Thus, the weight of the whole plate must be

$$\text{weight} = D \cdot H \cdot \int_0^W \rho(x)\,dx.$$

It isn't necessary to restrict our attention to rectangular plates. We can assume that the "profile" of our plate is the graph of any function f from a to b, as indicated at the left of the figure below. In this case, the density

throughout the shaded portion shown on the right is very close to $\rho(x_i)$, while the height of this portion is close to $f(x_i)$. So the weight of the shaded portion is close to

$$D \cdot f(x_i) \cdot (t_i - t_{i-1}) \cdot \rho(x_i) = D \cdot \rho(x_i) f(x_i) \Delta x_i.$$

This means that the weight of the whole plate is close to

$$D \cdot \sum_{i=1}^{n} \rho(x_i) f(x_i) \Delta x_i,$$

and consequently we must have

$$\text{weight} = D \cdot \int_{a}^{b} \rho(x) f(x) \, dx.$$

In general, integrals will be involved whenever we have to determine any physical quantity in terms of its "density" throughout a body. This is hardly surprising, since, roughly speaking, integration amounts to summing over an extremely large number of extremely small elements, within which we can pretend that everything is constant.

There are also a number of purely mathematical concepts whose values can be expressed in terms of integrals. Some calculus courses may devote quite a bit of time to these formulas, because they provide further opportunities to hone the skills of evaluating integrals. Other courses, with a different emphasis, may mention them only in passing. True to the spirit of this journey, we aren't going to worry about calculations, but merely present the formulas themselves, as final souvenirs.

1. Volumes of revolution.   In the figure below, we have taken the graph of the function f, between a and b, and rotated it about the x-axis, to obtain

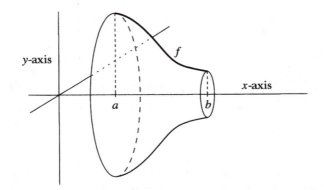

a *solid of revolution*. Even though the integral was originally invoked in order to calculate areas, we will be able to use it to determine the volume of this solid.

We begin, as usual, by choosing

$$a = t_0, t_1, \ldots, t_n = b,$$

together with points x_i between t_{i-1} and t_i, for each i. The figure on the left shows the portion of our solid between the planes $x = t_{i-1}$ and $x = t_i$. In the figure on the right we have approximated this portion by

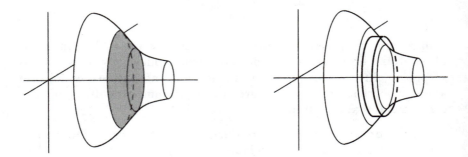

a cylinder between these same two planes, with the radius of the cylinder simply being $f(x_i)$. The volume of this cylinder is

$$\pi [f(x_i)]^2 \cdot (t_i - t_{i-1}) = \pi [f(x_i)]^2 \Delta x_i.$$

We can thus approximate our solid by a collection of cylinders, whose total volume is

$$\pi \cdot \sum_{i=1}^{n} [f(x_i)]^2 \Delta x_i.$$

So the volume of our solid of revolution must be the number that these sums are close to when all the Δx_i are very small, namely

$$\text{Volume} = \pi \int_{a}^{b} [f(x)]^2 \, dx.$$

In this example we obtained a sum of the form (∗) on page 113 with little work, but from now on things will not be so simple!

2. The shell game. In the following figure the area beneath the graph of the function f from a to b has been shaded. If this shaded area is rotated

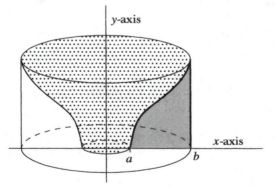

about the y-axis, instead of the x-axis, we obtain a more complicated solid: this solid is obtained by starting with the cylinder of radius b and then *removing* the dotted solid.

It might seem that it should be easy to calculate the volume of that dotted solid. After all, if we interchange the names of the x- and y-axes, then this solid is just one of the sort considered previously, except that instead of the function f we have to consider the function g whose graph is the graph of f rotated by 90°. But we may not be able to figure out a formula for this function g, and even if we can, the resulting integral might be much harder to calculate. Moreover, we might have a situation like that shown below (only the back half of the full rotation solid is shown).

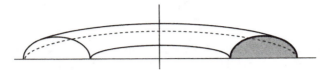

If we tried to simply interchange the x- and y-axes here, we would have to deal with a curve that isn't the graph of a function, so that we would have to break up the region into two or more pieces. The "shell method" is designed to deal with this problem.

The figure on the left shows (the back half of) a portion of our solid, the ring or "shell" that we obtain when we revolve the part of the graph from t_{i-1} to t_i. In the figure on the right we have approximated this shell

by a cylindrical shell whose height is simply $f(x_i)$. The volume of this cylindrical shell is

$$\pi \cdot f(x_i) \cdot [(t_i)^2 - (t_{i-1})^2] = \pi \cdot f(x_i)(t_i + t_{i-1})(t_i - t_{i-1})$$
$$= \pi \cdot f(x_i)(t_i + t_{i-1})\Delta x_i.$$

We can thus approximate our solid by a collection of cylindrical shells, whose total volume is

$$\pi \cdot \sum_{i=1}^{n} f(x_i)(t_i + t_{i-1})\Delta x_i.$$

Now these sums *aren't* of the form (∗) on page 113 for some new function g, so we will have to work a bit harder. The first thing we might notice is that our sum will look a lot better if we write it as two sums,

$$\pi \cdot \sum_{i=1}^{n} f(x_i)t_i \Delta x_i + \pi \cdot \sum_{i=1}^{n} f(x_i)t_{i-1}\Delta x_i.$$

If we were to choose

$$x_i = t_i \qquad \text{for each } i,$$

then the first sum would be of the desired form (∗), for the function $g(x) = xf(x)$. Similarly, if we were to choose all $x_i = t_{i-1}$, then the second sum would be of this form. Unfortunately, we can't choose the x_i to be both!

But since we are only interested in what happens when all Δx_i are small, i.e., when each t_{i-1} is very close to the corresponding t_i, you might expect that this shouldn't really matter, so that we should have the formula

$$\text{Volume} = 2\pi \int_a^b xf(x)\,dx.$$

This formula is indeed correct, although some nasty details must be handled to establish it rigorously.

3. Length. In the following figure we have used the points $a = t_0, t_1, \ldots,$ $t_n = b$ to produce a polygonal line, made up of the straight line segments

$$L_i \text{ from } (t_{i-1}, f(t_{i-1})) \text{ to } (t_i, f(t_i)).$$

The length of the graph of f from $(a, f(a))$ to $(b, f(b))$ should be the

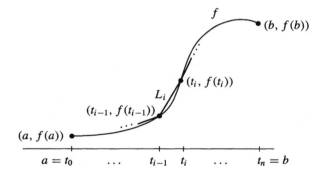

number that the lengths of such polygonal lines come close to when we make all Δx_i small.

You might think that length should be easier to compute than area, but it actually involves much more complicated formulas, because even the formula for the length of the segment L_i is more complicated:

$$\text{lengh of } L_i = \sqrt{(t_i - t_{i-1})^2 + \left[f(t_i) - f(t_{i-1})\right]^2}$$

$$= \sqrt{1 + \left[\frac{f(t_i) - f(t_{i-1})}{t_i - t_{i-1}}\right]^2} \cdot (t_i - t_{i-1}).$$

In order to do anything with this expression we need to use the Mean Value Theorem (yes!, the Mean Value Theorem on page 84, from Chapter 11, which you thought was only for theoretical mathematics geeks). This allows us to write

$$\sqrt{1+\left[\frac{f(t_i)-f(t_{i-1})}{t_i-t_{i-1}}\right]^2} = \sqrt{1+\left[f'(x_i)\right]^2} \qquad \text{for some } x_i \text{ with } t_{i-1} < x_i < t_i.$$

Consequently, the whole polygonal curve has length

$$\sum_{i=1}^{n}\sqrt{1+\left[f'(x_i)\right]^2}(t_i - t_{i-1}) = \sum_{i=1}^{n}\sqrt{1+\left[f'(x_i)\right]^2}\Delta x_i,$$

which is a sum of the form (∗) on page 113. Thus, the length of the curved graph must be given by

$$\text{Length} = \int_a^b \sqrt{1+\left[f'(x)\right]^2}\,dx.$$

4. Surfaces of revolution. Our last example combines the difficulties of both **2.** and **3.** and some additional preliminaries are also needed!

The figure on the left shows the surface of a pyramid made up of (ten) congruent isosceles triangles; the shaded area, between two parallel planes, is made up of congruent trapezoids, one of which is shown in the figure on the right. If this trapezoid has height h and bases l_1

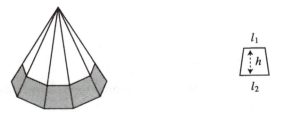

and l_2, then its area is $h(l_1 + l_2)/2$. So the area of the shaded region can be written as

(∗)
$$\text{area} = \frac{h(p_1 + p_2)}{2},$$

where p_1 is the perimeter of the top of the region and p_2 is the perimeter of the bottom.

A pyramid of this sort with very many sides will be close to a cone, with the shaded area being close to that shown below. If the top circle of this

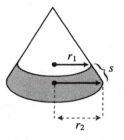

region has radius r_1 and the bottom circle has radius r_2, while s is the "slant-height" of this region, then $(*)$ shows that the area must be given by

$$\text{area} = \frac{s(2\pi r_1 + 2\pi r_2)}{2} = \pi s(r_1 + r_2).$$

The surface of the volume on page 116 can now be approximated by such conical pieces. For the piece from t_{i-1} to t_i, the radii are $f(t_{i-1})$

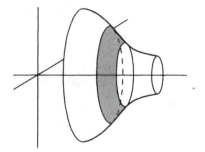

and $f(t_i)$ and its area is

$$\pi[f(t_{i-1}) + f(t_i)] \cdot s = \pi[f(t_{i-1}) + f(t_i)] \sqrt{(t_i - t_{i-1})^2 + \left[f(t_i) - f(t_{i-1})\right]^2}.$$

As before, we use the Mean Value Theorem to write the total of these areas as

$$\pi \sum_{i=1}^{n} [f(t_{i-1}) + f(t_i)] \sqrt{1 + \left[f'(x_i)\right]^2} \, \Delta x_i,$$

and then finally we use the additional argument from **2.** to conclude that

$$\text{Surface area} = 2\pi \int_a^b f(x) \sqrt{1 + \left[f'(x)\right]^2} \, dx.$$